This book surveys the recent advances in the field of ultrafast fiber switching devices and systems that have potential processing speeds in excess of 50 gigabits per second. The treatment is cross-disciplinary, covering basic physical principles, device physics and systems applications.

The purpose of the book is to provide a complete tour of the key research issues and approaches in ultrafast switching. A variety of all-optical devices are described and compared, and topics covered include routing and logic devices, solitons in optical fibers, and the application of ultrafast gates to telecommunications transmission systems, local area networks and optical computing. In highlighting the key research issues, the book also provides a perspective on future technological challenges. The author is a member of the Technical Staff in basic research at AT&T Bell Laboratories, Holmdel, and is excellently positioned to provide this stimulating survey of the state-of-the-art in this exciting and advancing field.

The book will be of great value to all researchers and graduate students working in the areas of nonlinear guided-wave optics, photonic switching, optical data processing and high-speed electronics.

T0275603

CAMBRIDGE STUDIES IN MODERN OPTICS: 12

Series Editors
P. L. KNIGHT
Optics Section, Imperial College of Science, Technology and Medicine
W. J. FIRTH
Department of Physics, University of Strathclyde

Ultrafast fiber switching devices and systems

TITLES IN THIS SERIES

Ultrafast fiber switching devices and systems

MOHAMMED N. ISLAM

AT&T Bell Laboratories Division

CAMBRIDGE UNIVERSITY PRESS
Cambridge, New York, Melbourne, Madrid, Cape Town, Singapore, São Paulo

Cambridge University Press
The Edinburgh Building, Cambridge CB2 2RU, UK

Published in the United States of America by Cambridge University Press, New York

www.cambridge.org
Information on this title: www.cambridge.org/9780521431910

First published 1992
This digitally printed first paperback version 2006

A catalogue record for this publication is available from the British Library

ISBN-13 978-0-521-43191-0 hardback
ISBN-10 0-521-43191-3 hardback

ISBN-13 978-0-521-02590-4 paperback
ISBN-10 0-521-02590-7 paperback

This book is dedicated to my parents, whose support and inspiration have always been with me

Contents

Contents

Preface

This book emphasizes device physics, but making novel and useful devices requires both a knowledge of basic physical principles and an understanding of system applications. The basic physical principles identify the fundamental limitations and provide the "bag of tricks" for implementing various functions, and the system determines the specifications and properties that a useful device must have. Complexity is rarely reduced but is simply moved from one place to another. The goal of clever device design is to move the complexity to where it can most easily be handled. By considering the entire process, we can understand where the complexity has been transferred and the price and benefits of a particular device design.

I have chosen examples that offer a complete tour of key research issues and approaches. My starting point is an explanation of the way that ultrafast switching complements existing technology and research, moving then to identifying the requirements and difficulties of implementing an ultrafast switch. I have then examined various physical phenomena and selected the principles that might help to tackle these difficulties. In a properly chosen configuration, the physical effects can be harnessed to implement a device. To extend and improve the device, a model must be developed. We also need a logical abstraction as a design tool for device-independent system architectures. Finally, various system applications are included to identify the opportunities and challenges provided by the novel devices. These applications relate back to the original motivations, thus closing the research loop.

The topic of this book is ultrafast, time-domain switching devices and systems. I concentrate on routing and logical switches that have the potential of operating at speeds greater than 50 gigabits per second, which is beyond the limits where electronics might be expected to operate. Also, I restrict the discussion to nonlinear guided-wave devices in glass or optical fibers that can be used for serial processing. The applications presented are pipelined, feed-forward designs that assume high-speed, long-latency gates. Ultrafast switching is a rapidly advancing field, and the examples presented are at an early stage of research. Therefore, the details of the examples are not as crucial as the issues and concepts that they illustrate. Also, this book is not a comprehensive review of the field; instead, the examples are chosen to illustrate various approaches. Since my personal research centers on ultrafast soliton switching in fibers, many of the examples are selected from soliton switching. For terabit-rate switches that use picosecond or femtosecond optical pulses in fibers, I believe that solitons are the natural data bits.

I hope that this book will appeal to applied physicists and engineers alike. The primary audience will probably be researchers and graduate students working in the areas of nonlinear guided-wave optics, photonic switching and optical data processing and high-speed electronics. Chapter 1 focuses on the motivation and research issues in ultrafast switching. Chapters 2 and 3 cover routing and logical devices and have an applied physics flavor. Timing and systems applications are described in Chapters 4 and 5, which have more of an engineering flavor. Chapter 6 summarizes the topics covered in the book and looks toward the future technological challenges for ultrafast switching. Although mentioned throughout the book, the basic physical principles and mathematical treatments of solitons are collected and detailed in the Appendices. I hope that the material shows not only the experimental and theoretical details, but also a research approach that addresses how to choose relevant problems, isolate key issues and move to the next stage of questions.

The exercise of writing this book has served as a vehicle for organizing my own thoughts on where the field of ultrafast switching is heading. For my current perspective I must acknowledge collaborations and discussions with many people including Erich P. Ippen, Curtis R. Menyuk, Linn F. Mollenauer, Jon R. Sauer and Roger H. Stolen. Special thanks to Michael W. Chbat, Chien-Jen Chen and Donna C. Cunningham for reading and commenting on the entire manuscript. David A. B. Miller has sharpened my focus on the key issues in photonic switching and also gave helpful suggestions on Chapters 1 and 6. James P. Gordon has been my "guru" on soliton physics and commented extensively on the Appendices. I especially thank Carl E. Soccolich, with whom I have worked closely over the past few years. Finally, I thank many coworkers in the field who have contributed and granted permission to reproduce some of their figures.

Abbreviations

AOM	acousto-optic modulator
BS	beam splitter
C	control
CCL	color center laser
CLK	clock
CNLS	coupled nonlinear Schrödinger equations
cw	continuous wave
DBS	dichroic beam splitter
DS	dispersion shifted
DWDM	dense wavelength division multiplexed
ECL	external cavity laser
EDFA	erbium-doped fiber amplifier
4WM	four-wave-mixing
GEO	generalized exclusive-OR
LD	laser diode
MBF	moderately birefringent fiber
MI	modulational instability
MIPRS	modulational instability polarization rotation switch
MOD	modulator
NLDC	nonlinear directional coupler
NLSE	nonlinear Schrödinger equation
NOLM	nonlinear optical loop mirror
PC	polarization controller
PBS	polarizing beam splitter
PM	polarization maintaining
POL	polarizer
PZT	piezo-electric transducer
S	signal
SCL	semiconductor laser
SIG	soliton-interaction gate

SDLG	soliton-dragging logic gate
SSFS	soliton self-frequency shift
STAG	soliton-trapping AND-gate
TDCS	time-domain chirp switch
TSI	time-slot interchanger
XOR	exclusive-OR

1

Introduction

Future switching systems are expected to process net data rates approaching a terabit per second (Tbit/s). The terabit benchmark is significant from a research standpoint because it means that the system will require different devices and architectures than are currently in use. These future systems may use some aspect of photonic switching to take advantage of inherent optical properties. For example [1.1], optics can be used beneficially in: (1) photonic interconnections because optics provides a quantum impedance transformation at a detector; (2) highly parallel logic operations because in free space light beams can pass through each other without interference; and (3) ultrafast switching devices because of the instantaneous nature of virtual optical transitions. In the first two applications terabit throughputs can be achieved by using massively parallel arrays of opto-electronic devices operating at megahertz speeds. In contrast, the third application is serial in nature and must use devices with speeds approaching a terabit per second. Applications in which serial devices will be important include high-performance front and back ends of telecommunications systems as well as fiber local area networks. The interest in such ultrafast switches stems from their ability to answer two questions. First, how can processing beyond electronic speeds be accomplished? Second, how can the bandwidth-rich environment provided by optical fibers be further utilized? For instance, in the low-loss window between 1.3 and $1.6\,\mu$m there is about 40 THz of bandwidth, and to exploit this advantage for time-division-multiplexed systems requires ultrafast switches. One goal of ultrafast switching is to make bandwidth an inexpensive, virtually limitless commodity in the system.

In this book I focus on ultrafast, serial, nonlinear guided-wave switches and their potential systems applications. "Ultrafast" means having speeds greater than 50 Gbit/s, or at least beyond the speeds that electronic systems may reach. For example, P. W. Smith [1.2] predicts that VLSI-like electronics may be limited above ∼ 35 GHz because of fundamental considerations such as transit, relaxation and thermal diffusion times. In addition, I restrict the discussion to devices based on the third-order nonlinear susceptibility $\chi^{(3)}$, so the output frequency can equal at least one of the input frequencies. Although the second-order susceptibility $\chi^{(2)}$ is stronger than $\chi^{(3)}$ in materials without inversion symmetry, the output is at the sum or difference frequency of the two inputs. As a result, the system becomes more complicated because cascading gates require additional devices for wavelength up- and down-conversion.

Devices that are based on "all-optical" interactions rely on virtual transitions in the material: i.e., the interaction is through deformation of wave functions, which is non-resonant and can be almost instantaneous. Since electrons are not "created," the devices are not limited by carrier recombination times in the material. In general, all-optical switching can be realized well below the bandgap of materials, thereby avoiding linear and nonlinear absorption and the related heating effects that can be detrimental at high bit rates. For example, optical fibers are typically used below one-fifth of the energy gap and semiconductors may be used below their half-gap energy. Furthermore, unlike electronic devices where the energy incident on the device leads to heating, most of the energy incident on the waveguide or fiber devices is guided and reappears at the output of the device.

Ultrafast devices can be divided into two general categories that are illustrated in Fig. 1.1. The first is a routing switch in which the input is connected to one of several output ports, and the routing is based on either the intensity of the signals or an externally supplied control beam. If only one output port is employed, then the routing switch works like an on–off switch. Also, if the routing is based on the intensity of the input, then the device may be used as a limiter or a saturable absorber. Routing switches are "physical" switches since photons are physically moved from one port to another. The other category is a logic gate (Fig. 1.1(b)) in which a Boolean operation is performed based on the values of the input signals. The logical approach can be powerful because it allows intelligence to be distributed throughout the system (in the sense that one data stream can control another), and this is one reason that modern electronic systems operate based on digital logic.

ROUTING SWITCH

LOGIC GATES

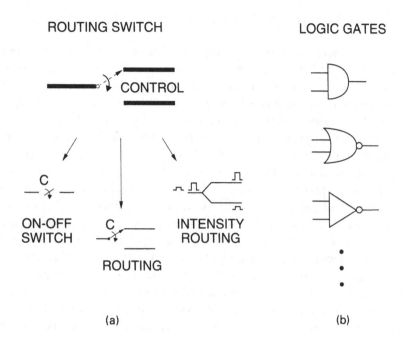

(a)

(b)

Fig. 1.1 (a) Routing switch in which the input is connected to one of several output ports. (b) Logic gate in which a Boolean operation is performed based on the values of the input signals.

Routing and logic switches differ fundamentally in the manner of the control. In routing switches, the control is typically in a different physical format than the data, and the control network may be external to the switching fabric. In a logic gate, on the other hand, the control is in the same physical format as the data, and, therefore, the control can be distributed throughout the switching fabric. Another difference between the two devices of Fig. 1.1 is the representation of the decision. A routing switch represents its decision by the position or location of the data, while the output of a logic gate has a "0" or "1" logic level. Since routing switches route the same photons from the input to the output, the signals may degrade because of loss, dispersion or cross-talk. In digital logic gates, the signal level and timing is regenerated at the output of each gate by replacing the input photons with new photons from a local power supply. The penalty for high-speed, digital-logic-based systems is that the switching energy and power supply requirements are major constraints.

Challenges for all-optical switching devices

There are many challenges in implementing devices that rely on all-optical interactions, and listed here are some of the key issues. First, how can we make three-terminal devices whose behavior is independent of the phase between signals? In general, the two input signals to a device may originate at different points in the system and their relative phase will be arbitrary. All-optical interactions are coherent processes in which the input signal phases are preserved. Absorptive opto-electronic devices, on the other hand, are based on incoherent processes because the incident light beams generate current, and the phase of the electrons are randomized from collisions with phonons in the material. Phase insensitivity restricts the useful nonlinear processes to those that depend on the intensity but not the electric fields of the inputs. Second, how do we make cascadable gates where the output looks like the input? Except in a few trivial situations, to complete a useful operation requires several decisions and corresponding levels of logic, so we must have devices that can drive similar devices. Although it is conceivable that devices in each level of logic can be tailored for the given input conditions, the simplest system involves using the same devices throughout. Third, how do we achieve small-signal gain where a smaller signal controls a larger pulse? Devices, in general, need gain to fan-out to several devices and to compensate for system losses in connecting to the next gate. The splitting and coupling losses between stages can be partially compensated for by introducing amplifiers, but at the cost of increased system-hardware complexity and additional spontaneous noise. Fourth, how can we lower the switching energy without a large increase in the device size? At this point the fundamental problem of potential terabit systems appears to be power supply limitations. For example, the average power required for a gate is the switching energy times the bit rate, which means that to switch a picojoule energy device at a terabit rate requires a laser with a watt of average power. The lasers must provide this large power at high repetition rates, in short pulses and at wavelengths compatible with the remainder of the system. Finally, how can we handle the timing constraints and synchronization required for terabit systems? As the bit rate increases, the bit period decreases and the tolerance to timing jitter and clock skew decreases. In a synchronous system all parts of the system must be phase and frequency locked to a master clock, and timing jitter must be minimized at each stage to avoid accumulation of errors. This book describes devices that address each of these issues.

Aside from the device issues, there are also critical materials considerations for implementing all-optical devices. First, we need to know the strength of the interaction, which for refractive index devices is proportional to the nonlinear index n_2: this determines the switching energy, which is typically inversely proportional to the product of n_2 and the length of the device. Second, we need to know if the pulse is distorted while propagating through the device. Sources of pulse distortion include nonlinear absorption and low-frequency Raman effects. A rule of thumb that summarizes the requirements on the real and imaginary parts of the third-order susceptibility $\chi^{(3)}$ in all-optical materials is that the nonlinear interaction must lead to a π-phase shift with less than 3 dB of absorption [1.3, 1.4]. A π-phase shift is required to switch from constructive to destructive interference (e.g., in a Mach–Zehnder interferometer); also, for intensity-dependent switching the interaction slows down or turns off once the amplitude has dropped to about half of its input value.

Two material systems have been demonstrated to satisfy the above rule of thumb for all-optical switching. Semiconductors can be used below half of the energy band gap and can lead to compact, integrable devices [1.5–1.7]. However, most of the all-optical devices described in this book have been demonstrated in optical fibers. Although the nonlinearity in fused silica fibers is weak, the very low loss and excellent guiding properties of optical fibers mean that long interaction lengths can be employed. In addition, fused silica has for all practical purposes an instantaneous response time since it is generally used far off resonance. For both of these reasons, fibers turn out to have figures of merit several orders of magnitude higher than almost any other nonlinear material [1.8], where the figure of merit is typically proportional to the nonlinear coefficient and inversely proportional to both the absorption coefficient (linear and nonlinear) and the response time. Fibers also have a mature fabrication technology, and the equations governing their behavior are well understood. Therefore, although fiber devices may have a long latency (delay from the input to the output), they at least permit exploration of various switch architectures.

In addition to using fibers, many of the interesting devices described in this book use solitons. Solitons can be defined as pulses that propagate nearly distortion-free for long distances in fibers (for a review of soliton properties see Appendix A). They occur in the anomalous group-velocity dispersion regime of fibers (e.g.,

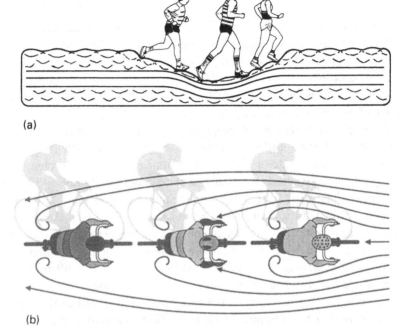

(a)

(b)

Fig. 1.2 (a) A simple analogy of runners on a mattress to intuitively explain soliton formation in fibers. The runners create a moving valley that pulls along the slower runners and retards the faster ones [1.9]. (b) Another analogy for solitons in which a group of bikers travel into the wind and draft-off one another.

wavelengths longer than $1.3\,\mu m$), and solitons represent a balance between the nonlinearity and the dispersion in the fiber.

To understand intuitively soliton formation in an optical fiber, Evangelides [1.9] has proposed a simple analogy of runners on a mattress. Figure 1.2(a) shows that the runners create a moving valley that pulls along the slower runners and retards the faster ones. In an optical fiber, the high-intensity pulse creates a valley of higher index-of-refraction material, which accelerates the slower low-frequency components and retards the faster high-frequency ones. Another analogy (Fig. 1.2(b)) is to consider a group of bikers traveling into the wind and drafting off one another (i.e. riding close together to cut wind resistance for those behind the leader). Normally the bikers would spread because of their different peddling speeds. However, the wind resistance impedes the fast bikers while the slower bikers have their headwind broken by the front bikers. Consequently, the slow and fast bikers travel as a packet.

The key feature of solitons that is used for switching applications is that they act in many ways like fundamental data bits and the entire pulse switches as a unit [1.10]. For example, pulse shape distortions generally accompany intensity-dependent switching of hyperbolic secant or Gaussian pulses: i.e., these pulses have a continuous range of intensities and each value of intensity switches differently. On the other hand, a fundamental soliton has a uniform phase across the pulse and undergoes complete pulse switching, which is important for obtaining cascadable gates with high contrast ratio. In addition, because solitons are balanced by two counteracting forces during propagation, they are stable against many perturbations such as birefringence or polarization dispersion [1.11]. Furthermore, fundamental solitons try to maintain a constant-area pulse, which means that after a soliton travels through an amplifier both its amplitude and its pulse shape can be restored [1.12].

There are several reasons why solitons are particularly advantageous for time-domain, terabit-rate switching using picosecond or femtosecond pulses in optical fibers. First, ultrashort pulses with wide bandwidths are generally adversely affected by group-velocity dispersion and nonlinearity in the fiber; but, in a soliton these two deleterious affects are kept in balance. For example, the characteristic length for a soliton is called the soliton period Z_0. Without the nonlinearity, the pulse would begin to spread due to dispersion at a distance Z_0, and beyond that would spread in proportion to Z/Z_0 (a sech t input broadens asymptotically toward $\text{sech}(tZ_0/Z)$ without nonlinearity). The soliton period is defined as $Z_0 = 0.322\pi^2 c\tau^2/\lambda_0^2 D$, where D is the group-velocity dispersion and τ the pulse width, and for $\tau \sim 0.5\,\text{ps}$ pulses, Z_0 may range from one to ten meters. Therefore, for any reasonable size fiber system or fiber logic gate we must use solitons. Second, all-optical switching can take advantage of several unique properties of solitons, including modulational instability, elastic collisions, and soliton dragging and trapping, and these phenomena will be discussed in detail in Chapters 2 and 3. Third, as will be elucidated in the time-domain chirp switch architecture (Section 3.2), the particle-like nature of solitons can lead to low energy switching since a small frequency change can lead to a large time shift. Finally, solitons are now being seriously studied for use in long-haul telecommunications systems. Therefore, soliton switching aligns our switching effort with long-haul transmission and enables us to take advantage of technological developments like erbium-doped fiber amplifiers.

This book is organized as follows. Chapter 2 covers a variety of routing switches including Kerr gates, four-wave-mixing gates, non-linear directional couplers and Mach–Zehnder interferometers. Chapter 3 concentrates on digital logic gates that are based on solitons in optical fibers, such as soliton-dragging logic gates, soliton-trapping AND-gates and soliton-interaction gates. Then, timing considerations and techniques for handling jitter in terabit systems are described in Chapter 4. Chapter 5 looks at potential applications of ultrafast devices for serial processing in telecommunications and optical computing. Finally, Chapter 6 summarizes the book and looks at future prospects for ultrafast switching. The Appendices review the basic properties of solitons that are relevant to ultrafast switching. Appendix A treats the single-axis soliton, Appendix B extends the treatment to birefringent fibers, and Appendix C treats different frequency solitons along the same axis. It is worth reiterating that most of the devices presented are at an early stage of research, and they should be viewed as demonstrations of principle. Furthermore, only a few examples of devices and applications are given, to illustrate the concepts and issues. This book is by no means an exhaustive review of all-optical switching.

2

Routing switches

Switches based on the nonlinear index of refraction n_2 that route their input to one of several outputs (Fig. 1.1a) are perhaps the earliest and most widely studied all-optical gates. Five examples – Kerr gates, four-wave-mixing gates, nonlinear directional couplers, Mach–Zehnder interferometers and Manchester-coded solitons – will serve to illustrate various concepts. Kerr gates rely on n_2 and intensity-dependent switching induced by a control beam that is generally at a different optical wavelength. As a single input and output device, this can be used as an optical limiter. As a three-terminal device it does not have gain since a larger pulse is required to control a smaller pulse.

Four-wave-mixing gates also rely primarily on n_2, but additionally require some sort of phase matching between the two inputs. The phase matching leads to parametric gain so that a small signal can control a larger signal, which corresponds to fan-out or small-signal gain for the device. The contrast ratio for a four-wave-mixing gate can be large if we monitor only the new frequencies that are generated through the mixing. When operating in the soliton regime of fibers, the nonlinearity can participate in the phase matching for four-wave-mixing, which leads to devices based on modulational instability.

Nonlinear directional couplers (NLDCs) are typically used as a single input, intensity-dependent routing switch. Although NLDCs exhibit intriguing physics, their only use in systems may be as a saturable absorber to provide system clean-up. NLDCs are dual-mode devices based on two coupled waveguides, such as dual-core fibers or the two polarization axes of a fiber, and switching is achieved by using the intensity-dependent index change to detune the waveguides and disrupt the coupling. NLDCs also nicely illustrate the pulse break-up problem of instantaneous nonlinearity switching devices and indicate the need for square pulses or solitons to achieve complete switching.

Another example of a two-mode system is a Mach–Zehnder inter-ferometer, which uses the phase difference between two arms to vary between constructive and destructive interference. A stable fiber implementation of the interferometers is a nonlinear optical loop mirror (NOLM) or a nonlinear Sagnac device. Solitons have been used in nonlinear loop mirrors to demonstrate complete pulse switching and to achieve relatively low switching energies because of the long fiber lengths used. A routing switch can be made by adding a control beam at a different frequency, and logic operations can also be obtained by crossing fiber axes and having orthogonally polarized pulses repeatedly interact through cross-phase modulation.

An ultrafast, all-optical "2-module" routing gate, which is a simple and complete building block for an extended generalized shuffle net-work, uses Manchester-coded soliton pulses and behaves like a polari-zation rotation switch with fan-out. In Manchester coding, a soliton with a falling slope at a reference time corresponds to a "0", and a soliton with a rising slope corresponds to a "1". The temporal shifting of pulses used by this 2-module switch serves as a lead into the time-shift keyed data format that is described in Chapter 3.

2.1 Kerr gates

A Kerr modulator uses the change in polarization state that is due to the intensity-dependent refractive index n_2. Duguay and Hansen [2.1] first used a bulk nonlinear Kerr shutter for ultrafast sampling measurements, and Stolen, et al. [2.2], first demonstrated non-linear polarization rotation in a fiber. As a two-input device, a Kerr gate can be used as an ultrafast modulator [2.3] or an optical demulti-plexer [2.4]. If a single input is used with soliton pulses, then the Kerr gate can act as an intensity discriminator [2.2] or an optical lim-iter [2.5].

The typical configuration for a Kerr modulator is shown in Fig. 2.1, where a weak signal at frequency ω_2 is gated by a strong pump at ω_1. The strong pump is polarized along one axis of a polarization-maintaining fiber, while the weak signal is polarized at 45° to the axis. The frequency filter at the fiber output removes the pump at ω_1. The wave plates are adjusted so that the polarizer blocks the weak signal in the absence of the pump, and the pump increases the probe transmission through the optically induced birefringence. The power transmission through the polarizer is proportional to $\sin^2(\frac{1}{2}\Delta\phi)$, where

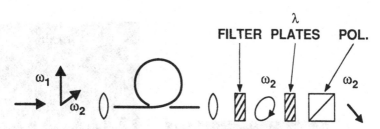

Fig. 2.1 Schematic of a fiber Kerr modulator. A strong optical pulse at ω_1 changes the polarization state of a signal at ω_2 by the intensity-dependent nonlinear index [2.32].

$$\Delta\phi = \frac{2\pi L}{\lambda}(\delta n_\parallel - \delta n_\perp); \qquad \delta n_\parallel - \delta n_\perp = n_2 I_p \tag{2.1}$$

and I_p is the pump intensity and L is the length of the fiber.

Morioka, et al. [2.4], have used the Kerr effect to demultiplex a train of 30 ps optical pulses from a gain-switched laser diode using control pulses from a mode-locked Nd:YAG laser. A schematic diagram of their Kerr demultiplexer set-up is shown in Fig. 2.2. Two problems of using long lengths of high birefringent fibers are: (1) polarization dispersion limits the effective interaction length for short pulses; and (2) the temperature-dependent birefringence causes polarization fluctuations of the signal pulse. Both of these limitations can be circumvented by canceling the overall birefringence by splicing together two

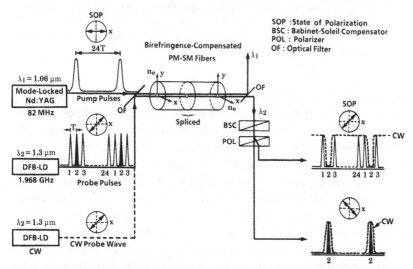

Fig. 2.2 Experimental set-up for the all-optical demultiplexer based on the optical Kerr effect in fibers [2.4].

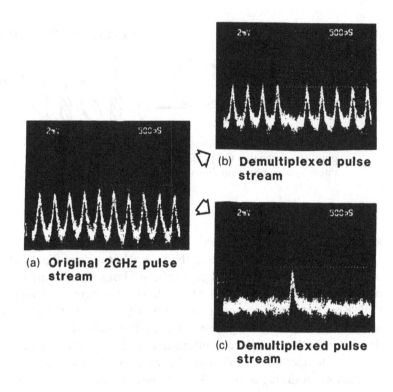

Fig. 2.3 Result of all-optical demultiplexing using the Kerr effect in fibers. (a) Original 2 GHz pulse stream. (b) Demultiplexed pulse stream in one arm. (c) Demultiplexed pulse stream in the other arm [2.4].

equal-length fibers with their axes crossed at right angles. For example, in Fig. 2.2 the Kerr medium consists of two 10 m lengths of polarization maintaining fiber spliced together with crossed axes. Synchronizing the 82 MHz repetition rate of the pump laser at 1.06 μm to every 24th pulse of the 1.3 μm probe laser diode (~2 GHz repetition rate) causes the pump to remove every 24th pulse. Figure 2.3 shows the demultiplexed probe pulses that are observed with a PIN photodiode. The cross-coupling from (b) to (c) in Fig. 2.3 was measured using a streak camera and found to be less than ~20 dB. Although the birefringence is compensated for in this experiment, the group velocity difference between the two optical pulses still limits the walk-off length. One solution is to choose the two wavelengths on opposite sides of the zero-dispersion wavelength of the fiber, so that the two group velocities are equal.

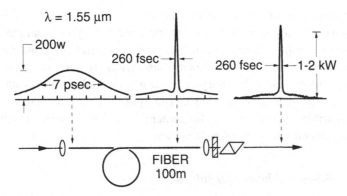

Fig. 2.4 A soliton pulse is compressed in a fiber, and then the uncompressed wings are removed by taking advantage of an intensity dependent state of polarization at the fiber output [2.6, 2.32].

Even without adding the control pulse, a self-induced birefringence in the fiber can be used to implement an intensity discriminator. If the two principal axes of a birefringent fiber are excited unequally, then a pulse has an intensity-dependent state of polarization [2.2]. The intensity-dependent change in beat length occurs because cross-phase modulation for orthogonally polarized pulses is only two-thirds the strength of self-phase modulation. The intensity dependent birefringence is given by $\delta n = \delta(n_x - n_y) = \frac{1}{3}n_2(I_x - I_y)$, where n_2 is the nonlinear index of the fiber and I_x and I_y are the intensities along the two principal axes. The polarization rotation corresponds to a transmission change when a polarizer is used at the fiber output.

Such an intensity discriminator can be used to separate the intense subpicosecond pulses formed by soliton compression from a weaker uncompressed pedestal [2.2, 2.6]. For this application the polarizer in Fig. 2.4 is adjusted to block the beam at low power, and the quarter-wave plate compensates for the intensity-independent birefringence. At high intensities, the state of polarization will rotate, and the perpendicular component will be transmitted by the polarizer. The power transmitted through the polarizer is given by

$$P_{\text{out}} = P_{\text{in}} \sin^2(\tfrac{1}{2}\Delta\phi)\sin^2 2\theta; \qquad \Delta\phi = \frac{2\pi L n_2}{3\lambda}(I_x - I_y) \qquad (2.2)$$

where $\theta = \tan^{-1}\sqrt{I_x/I_y}$ is the the angle between the polarization axis and the incident linear polarization.

As shown in Fig. 2.4, Mollenauer, et al. [2.6], have confirmed the pedestal removal experimentally by using the output of a color-center laser that was compressed in a fiber carrying solitons. The pedestal accompanying a 260 fs pulse is greatly reduced by using the nonlinear polarization rotation as a fast saturable absorber. This example for higher-order soliton compression illustrates a general principle found also in the normal group-velocity dispersion regime of a fiber. Namely, since a pulse has varying intensity, the polarization rotation is nonuniform over a pulse, thus causing pulse-shape distortions after the polarizer. However, a fundamental soliton shows a single intensity-dependent state of polarization.

All-optical limiter for solitons

The intensity-dependent polarization rotation of solitons in birefringent fibers can be used to make an ultrafast, all-optical limiter (i.e., a device whose transmission saturates at some intensity) [2.5]. Since the fundamental soliton has a single state of polarization, pulse-shape distortions after the polarizer are avoided. This limiting action could be useful to prevent the transfer of fluctuations from one stage to another in a switching or transmission system. The set-up for the limiter is the same as illustrated in Fig. 2.1, and for the limiter the polarizer is adjusted to maximize the transmission at low intensities.

To study the nonlinear polarization rotation for solitons, the coupled nonlinear Schrödinger equations (see Appendix B) were solved numerically in Ref. [2.5]. The coherence terms are neglected since the background and bend-induced birefringence are sufficiently large. In Fig. 2.5 we show the limiting action for $\theta = 42°$ and $L = 20Z_0$, where Z_0 is the soliton period. Polarization rotation begins at $0.8E_1$, where E_1 is the fundamental soliton energy, and the output energy is approximately flat between $1 < E/E_1 < 1.45$. Above this energy, the transmission reduces further and behaves like a sine-squared function. Although the clipped output energy varies by 4% between $1 < E/E_1 < 1.45$ and rolls over for $E/E_1 > 1.5$, a 45% energy range may be an acceptable window for guarding against unavoidable fluctuations.

An important point is that the limiting action does not lead to distortion of the output soliton. In Fig. 2.6 we show the input and output intensities at $E/E_1 = 1$, 1.21 and 1.44 for the limiter of Fig. 2.5. The output pulse shapes are identical to the input and have no features in the wings. The pulse widths change because fundamental solitons try to maintain a π-area pulse, as further discussed below. However, if the output pulses are allowed to propagate in another length of fiber, they will settle to nearly the same width as the

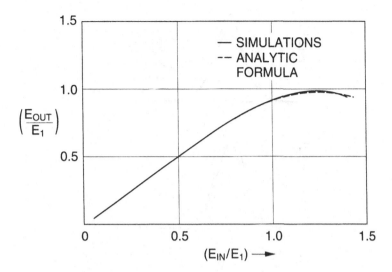

Fig. 2.5 Calculated optical limiter transfer function for solitons with $\theta = 42°$ and $L = 20Z_0$. The energies are normalized to the fundamental soliton energy E_1. The solid curve is from numerical simulations of the coupled nonlinear Schrödinger equation, and the dashed curve is from Eq. (2.3) and relation (2.4) with $\xi = 0.6$.

solitons readjust into π-area pulses. Moreover, it turns out that the phase of the limiter output is nearly uniform over the pulse width.

The transfer function in Fig. 2.5 can be approximated by simple formulas for polarization rotation including the asymptotic soliton solution. The transmitted power from the limiter for a continuous wave (cw) is the complement of relation (2.2) and is given by

$$P_{\text{out}} = P_{\text{in}}[1 - \sin^2(\tfrac{1}{2}\Delta\phi)\sin^2 2\theta];$$

$$\Delta\phi = \frac{\pi}{6}\frac{P}{P_1}\frac{L}{Z_0}\cos 2\theta, \qquad (2.3)$$

where P_1 is the fundamental soliton power and we have used soliton normalizations. For solitons we consider an input pulse of the form $u_i = (1+a)\operatorname{sech} t$ that contains a fundamental soliton for $-\tfrac{1}{2} < a < \tfrac{1}{2}$. As detailed in Appendix A, in this range of amplitudes a soliton that is along a single fiber axis reshapes to form a π-area pulse with an asymptotic soliton field $u_\infty = (1+2a)\operatorname{sech}[(1+2a)t]$. For $a < 0$ the pulse broadens in the fiber, and for $a > 0$ the pulse narrows. In this process, a fraction $[a/(1+a)]^2$ of the pulse energy is lost to dispersive waves. Although we now consider solitons propagating along two axes of the fiber, for the input u_i we try in Eq. (2.3)

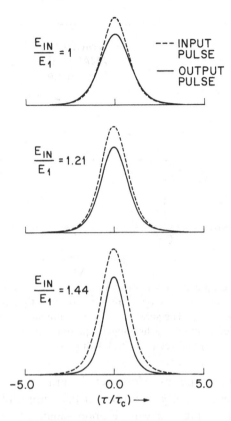

Fig. 2.6 Numerically calculated output intensities at $E/E_1 = 1$, 1.21, 1.44 for the limiter in Fig. 2.5.

$|u|^2 = P/P_1 \sim (1+2a)^2$. Note that the phase change $\Delta\phi$ versus amplitude is a sharper function for solitons than for ordinary pulses.

In addition to the asymptotic soliton amplitude, we must also distinguish a soliton from a cw. Although the fundamental soliton has varying intensity, it readjusts to have an average, uniform phase shift over the pulse. Therefore, we try the peak intensity of the soliton in Eq. (2.3) and include a correction factor ξ for the phase averaging; i.e.,

$$\Delta\phi_{\text{soliton}} \cong \frac{\pi}{6}\frac{L}{Z_0}\xi(1+2a)^2\cos 2\theta. \qquad (2.4)$$

When a fundamental soliton propagates along a single axis of a fiber, ξ is exactly $\frac{1}{2}$ for self-phase modulation. However, by fitting transfer functions for several θ values and $L/Z_0 = 5$, 10 and 20, we find numerically that $\xi \sim 0.6$ for the case with both self- and cross-phase modulation [2.5]. For example, the dashed curve in Fig. 2.5 corresponds to Eqs. (2.3) and (2.4) with $\xi = 0.6$.

In the experiments, $\tau \sim 260\,\text{fs}$ Gaussian pulses at $\lambda \sim 1.685\,\mu\text{m}$ are obtained from a multiple-quantum-well passively mode-locked NaCl color-center laser [2.7]. A continuously rotating 12 mm thick glass substrate is used at the fiber input to vary the coupling and, hence, the power level in the fiber. One detector at the fiber output detects scattered light from the output lens as a measure of the power in the fiber, and another detector is placed after the polarizer. The outputs from the two detectors are used as the $x-y$ inputs to an oscilloscope. To minimize polarization scrambling, the 35 m fiber ($\sim 35Z_0$) is loosely laid in a 40 cm diameter circular container, and the maximum power coupled into the fiber (average power $\sim 25\,\text{mW}$) is approximately the fundamental soliton power.

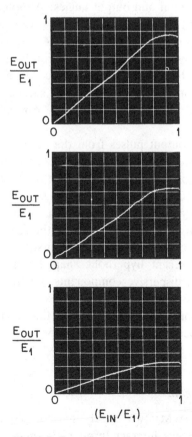

Fig. 2.7 Experimental transfer characteristics corresponding to limiting action for three sets of input and output angles.

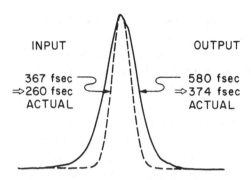

INPUT   OUTPUT

367 fsec   580 fsec
⇒260 fsec   ⇒374 fsec
ACTUAL   ACTUAL

Fig. 2.8 Autocorrelations of input and output pulses at the maximum input power for the top plot in Fig. 2.7.

We plot in Fig. 2.7 the transfer characteristics corresponding to limiter action for three sets of input and output angles. Although the precise conditions are different from Fig. 2.5, the qualitative behavior is verified. As can be seen from Eqs. (2.1.3) and (2.1.4), by varying the input angle θ, the power at which polarization rotation begins and the maximum transmission can be tailored. Furthermore, by changing the input and output angles, other transfer functions, such as high-intensity band-pass filters and intensity notch filters, can also be implemented.

The autocorrelations of the output pulses from the optical limiter confirm the absence of pulse shape distortions. For example, in Fig. 2.8 we show autocorrelations of the input and output pulses at the maximum input power for the top case in Fig. 2.7. The input pulse of 260 fs broadens to ~374 fs after the fiber and polarizer but is otherwise similar in shape. Although the results are close to the fundamental soliton power, broadening occurs for two reasons. First, the Gaussian input pulses evolve into secant hyperbolic shaped solitons by stripping off some non-soliton, dispersive components. Furthermore, a soliton propagating along two axes of a fiber requires more energy than for single axis propagation [2.8]. This confirms that nonlinear polarization rotation does not lead to pulse distortion for fundamental solitons.

2.2 Four-wave-mixing gates

In four-wave-mixing (4WM) two closely spaced frequencies from the control and signal beams interact through the nonlinearity $\chi^{(3)}$ to generate new Stokes and anti-Stokes sideband frequencies. If

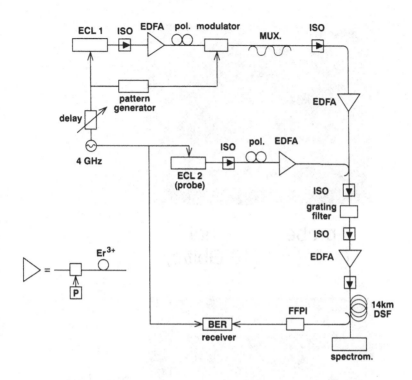

Fig. 2.9 Experimental configuration for testing an all-optical demultiplexer based on four-wave-mixing in a fiber. (ECL = external cavity laser, EDFA = erbium-doped fiber amplifier, pol. = polarization
controller, ISO = optical isolator, MUX = multiplexer, DSF =
dispersion-shifted fiber, FFPI = fiber Fabry–Perot interferometer,
BER = bit-error-rate, P = diode laser pump, spectrom. = scanning
spectrometer) [2.10].

we use a frequency filter to monitor only the newly generated wavelengths, then the output is effectively an AND-gate. Unlike self-phase
modulation or Kerr effects that are self-phase-matched, 4WM
requires phase-matching and temporal overlap between the control
and signal pulses to obtain parametric gain. Phase matching
(wavevector or momentum conservation) means that the Stokes and
anti-Stokes waves must together have the same magnitude of
wavevector as the sum of the wavevectors from the signal and control
wavelengths [2.9]. A simple analogy is a person who is pushing a
child on a swing. If the pushing action is in-phase with the swinging
motion of the child, then the amplitude of the child's motion

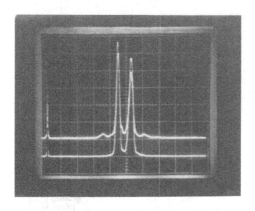

probe   signal
(4 GHz)   (16 Gbit/s)

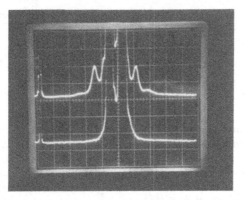

overlap

no overlap

Fig. 2.10 Signal and probe spectra measured at the fiber output with
and without pulse overlap in the fiber (top traces with pulse overlap;
bottom traces without pulse overlap). (top) Signal is 0.35 pJ and
probe is 0.82 pJ at the fiber input; (bottom) vertical scale expanded
by factor of 6 (horizontal scale = 1 nm/div, resolution 0.07nm, probe
repetition rate 4 GHz, signal at 16 Gbit/s) [2.10].

increases. On the other hand, if the push is at a different periodicity
than the swinging, then sometimes the push helps and sometimes it
hinders, so that on the average the child comes to a stop.

Andrekson, et al. [2.10], have demonstrated a high-speed optical
demultiplexer using control and signal pulses obtained from actively
mode-locked external cavity lasers (ECL). In the experimental
apparatus of Fig. 2.9, the ECL produce 20–25 ps pulses near 1.53 μm

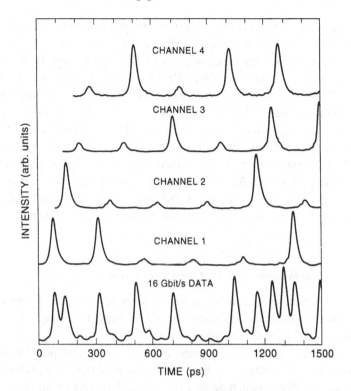

Fig. 2.11 Streak camera recordings of the output signals from the fiber 4WM demultiplexer after passing through a fiber Fabry–Perot filter. Demultiplexed data for channels 1–4 at 4WM component with data and probe pulses overlapping in the fiber are shown in the top four traces. The bottom trace shows a 16 Gbit/s data sequence [2.10].

and erbium-doped fiber amplifiers (EDFA) are used throughout the system to maintain sufficient power levels. The data is encoded with an external modulator and multiplexed up to a bit rate of 16 Gbit/s. The 4WM demultiplexer consists of a 14 km length of dispersion-shifted, single-mode fiber, and the control and signal beams are collinearly polarized using polarization controllers to maximize the interaction length. The wavelength separation between the two laser sources is 8 Å, which corresponds to a walk-off length of ∼ 500 km; thus, the phase matching between the signal and probe waves is satisfied.

The measured spectra of the signals after propagating through the fiber is shown in Fig. 2.10, and the sidebands generated by 4WM when the two pulses overlap are apparent. Therefore, 4WM is sensitive to the pulse overlap as well as the phase-matching condition. The conversion efficiency to the sideband frequencies is ~6%. Figure 2.11 shows streak camera recordings of the output signals detected with a 4 Gbit/s receiver after filtering in a narrow band fiber Fabry–Perot interferometer: the filter selects only one of the 4WM sideband frequencies. The bottom trace shows a multiplexed 16 Gbit/s sequence and the other traces show the demultiplexed data at the 4WM component for channel numbers 1–4. The apparent pulse broadening results from the relatively narrow bandpass (full width half maximum of 0.6 Å) of the Fabry–Perot filter. The extinction ratio in Fig. 2.11 is imperfect because the finite width of the Fabry–Perot pass band partially overlaps the control and signal wave-lengths.

Modulational instability fiber interferometer switch

When operating in the anomalous group-velocity regime of a fiber, the nonlinearity may enter the 4WM phase-matching condition to cause modulational instability (MI). In MI the amplitude and phase modulations of a wave grow because of the interplay between nonlinearity and anomalous dispersion. The equations governing MI, including the phase matching and frequency dependent gain relations, are derived in Appendix A. A modulational-instability-based fiber interferometer switch has been demonstrated [2.11] that operates near 1.5 μm and has gain from a small signal controlling a much larger input. In this device, pulses from a color-center laser (CCL) are gated through a fiber Mach–Zehnder interferometer by weak cw light from a semiconductor laser (SCL). The CCL pump beams ($P \sim 3$ W) enter both arms of the balanced interferometer, and the phases are adjusted to obtain a nulled output. When SCL power $\gtrsim$ 3 μW is injected into one arm, induced MI severely distorts the optical wave in that arm, thereby destroying the interferometer null. MI is an attractive amplifier because it relies on nonlinearities that are almost instantaneous, and it provides unidirectional gain so that feedback from the output is not so critical.

When two waves at frequencies ω_0 and $\omega_0 - 2\pi f$ (e.g., a pump and a weaker signal) propagate in a fiber, an amplitude modulation at the beat frequency f occurs. This amplitude modulation leads to a phase modulation through the intensity-dependent index. If the waves are in the anomalous group-velocity dispersion regime and the pump is sufficiently intense, then the field in each modulation period undergoes compression and develops into a narrow soliton pulse. As shown in Appendix A, the initial growth of a signal that is excited at the lower sideband is given by

$$P(\omega_0 - 2\pi f) = P_0\left\{1 + \left(\frac{A^2}{\gamma}\sinh\frac{\gamma l}{Z_c}\right)^2\right\}, \quad \gamma = \Omega\sqrt{A^2 - (\tfrac{1}{2}\Omega)^2},$$

$$(2.5)$$

where

$$\Omega = 2\pi f t_c, \quad z_c = 2\pi\frac{ct_c^2}{\lambda_0^2|D|}, \quad P_c = \frac{\lambda_0 A_{eff}}{2\pi n_2 z_c}, \quad A = \sqrt{\frac{P_p}{P_c}}$$

and P_p is the pump power. For typical numbers in the experiments ($\lambda_0 = 1.53\,\mu$m, $|D| = 4.13\,$ps/(nm km), $A_{eff} = 2.6\times10^{-7}\,$cm^2, $n_2 = 3.2\times10^{-16}\,$cm^2/W, $P_p \sim 3\,$W, $f = 280$ GHz and $l = 600\,$m) $2\gamma l/Z_c \sim 15.8$, which would mean a gain of 2.87×10^6. However, the gain saturates below 10^6 because of pump depletion and Raman effects or soliton self-frequency shift (SSFS) [2.12, 2.13].

A fiber Mach–Zehnder interferometer desensitizes the output to fluctuations in the pump, which equally affect both arms, while providing a sensitive measure of amplitude and phase changes in a single branch, which imbalances the interferometer. MI is induced in only one arm of the interferometer and the difference output $\tfrac{1}{2}|u_1 - u_2|^2$ is monitored, where u_1 and u_2 are the electric fields in the arms. The interferometer enhances the gain from MI since the output includes not only the power in the various sidebands, but also a large fraction of the pump power because of depletion and phase changes in the pump.

Experimentally, the two orthogonal axes of a single-mode, polarization-preserving, dispersion-shifted fiber (600 m long) are used as the two arms of the Mach–Zehnder interferometer. A $Tl^0(1)$:KCl CCL provides the 1.53 μm pump pulses for the apparatus in Fig. 2.12: pure cw pump is not used because of power requirements and to avoid backward-stimulated Brillouin scattering. An étalon in the CCL broadens the pulses to $\tau \gtrsim 50\,$ps (since $\tau \gg 1/f$, this serves as a quasi-cw pump). The CCL output is split into two equal parts, and one arm's polarization is rotated by 90°. After the two beams are combined with a polarizing beam splitter, another beam splitter couples in the signal or perturbation field from a single-mode distributed-feedback SCL with asymmetric coatings. The frequency of the SCL is $\sim 300\,$GHz shifted from the CCL center wavelength, and the frequency separation can be varied by temperature tuning the SCL. A

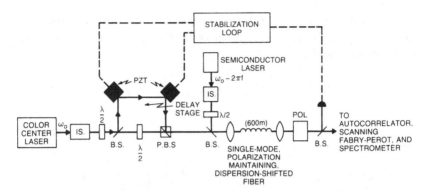

Fig. 2.12 Experimental apparatus for testing the modulational-instability-based fiber interferometer switch (IS = isolator, BS = beam splitter, PBS = polarizing beam splitter, POL = polarizer, PZT = piezo-electric transducer).

polarizer analyzes the fiber output and is set for a null when the SCL is absent.

When the interferometer is balanced and nulled with $\sim 3.5\,\mathrm{W}$ of peak pump power in each arm, the autocorrelation of Fig. 2.13(b) and the spectrum of Fig. 2.13(d) emerge after the polarizer. Although the broad pump pulse gives a low autocorrelation signal, the leakage is clear from the spectrum. The leakage power could be reduced by improving the temporal and spatial overlap at the analyzer and by using better polarizers.

When $4.4\,\mu\mathrm{W}$ of SCL power at $280\,\mathrm{GHz}$ below the pump frequency is injected into one polarization of the fiber, MI causes the autocorrelation and spectrum to change dramatically [Figs. 2.13(a) and 2.13(c)]. The net gain at the device output after the polarizer is $\sim 4.2 \times 10^4$ ["gain" being defined as the ratio of the peak power at the device output to the SCL input power, $G(\omega) = P_{\mathrm{out}}(\omega)/P_{\mathrm{in}}(\omega_0 - 2\pi f)$], while the gains at the various frequencies are $G(\omega_0) = 1.7 \times 10^4$, $G(\omega_0 - 2\pi f) = 10^4$ and $G(\omega_0 + 2\pi f) = 8.1 \times 10^3$. Despite these high gains, the on–off contrast ratio, which is the ratio of output energies with and without the SCL, is only $\sim 3{:}1$ because of the pump leakage. The change in autocorrelation is more pronounced than the spectral change because the autocorrelation is proportional to intensity squared rather than energy and because the second-harmonic process is strongly polarization sensitive.

The power output and contrast ratio improve when higher SCL powers are used. For example, with $\sim 3\,\mathrm{W}$ of CCL power in each arm and $58\,\mu\mathrm{W}$ of SCL power ($f = 275\,\mathrm{GHz}$), the measured gains

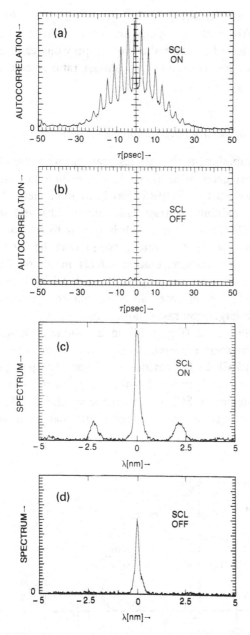

Fig. 2.13 (a), (b) Autocorrelations and (c), (d) spectra of modulational-instability-based fiber interferometer switch output with and without semiconductor laser (SCL) perturbation. The same scales are used for the two autocorrelations and the two spectra ($P_{ccl} \sim 3.5\,\mathrm{W}$ in each arm, $P_{scl} \sim 4.4\,\mu\mathrm{W}$, $f = 280\,\mathrm{GHz}$, 600 m of fiber).

after the polarizer are $G(\omega_0) = 4 \times 10^3$, $G(\omega_0 - 2\pi f) = 640$ and $G(\omega_0 + 2\pi f) = 380$. Although the gains are lower, the net output power is almost a factor of 2 larger than in the previous case. Since the leakage is more or less constant, the contrast ratio is now more than 5:1.

Modulational instability polarization rotation switch

In another embodiment that shares features with the above MI switch and the Kerr gates of Section 2.1, a modulational instability polarization rotation switch (MIPRS) has been implemented that operates near 1.5 μm, exhibits contrast ratios up to 40:1 and small-signal gains up to 40 dB [2.14]. This switch is similar to the modulation instability fiber interferometer switch, except that the interferometer is replaced by polarization rotation effects in fibers. In the MIPRS, pulses from a CCL are again gated through a length of fiber by low power, cw light signals from an SCL, but the polarization rotation leads to much larger contrast ratios at the output.

The device is implemented (Fig. 2.14) in a 500-meter length of single-mode, non-polarization preserving, dispersion shifted fiber, and a mode-locked $Tl^0(1){:}KCl$ CCL provides $\tau \approx 75$ ps, 1.54 μm pulses that serve as the "quasi-cw" pump for the experiment. A single-frequency distributed-feedback SCL that is frequency shifted 832 GHz from the CCL center frequency provides the signal that controls the

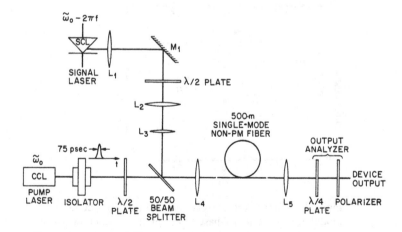

Fig. 2.14 Experimental apparatus for the modulational instability polarization rotation switch (L = lens, M = mirror, λ/2 = half-wave plate, λ/4 = quarter-wave plate).

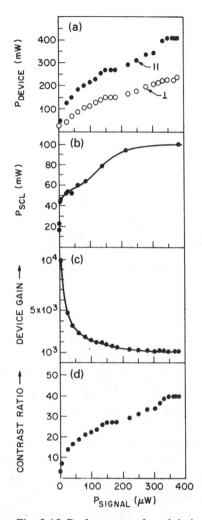

Fig. 2.15 Performance of modulational instability polarization rotation switch as a function of semiconductor laser (SCL) signal power
($P_{ccl} = 4.5\,\mathrm{W}$, $\tau \sim 75\,\mathrm{ps}$, $\Delta f = 832\,\mathrm{GHz}$). (a) Device output power
for signal and pump beams polarized parallel ($\|$) and perpendicular
($\perp$) at the fiber input. (b) SCL power at fiber output without polar-
izer (parallel case). (c) The net device gain (gain equals $P_{\mathrm{device}}/P_{\mathrm{scl}}$
in the fiber) and (d) contrast ratio ($P_{\mathrm{on}}/P_{\mathrm{off}}$) for the parallel case.

device output. A quarter waveplate is used at the fiber output to
transform the elliptically polarized pump beam back to a linear state
of polarization, which is then nulled using a polarizer.

Figure 2.15 summarizes the performance of the MIPRS as a function of signal SCL power for 4.5 W of peak pump power. In Figs.

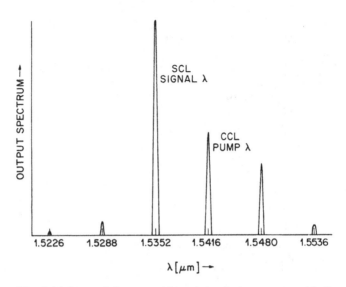

Fig. 2.16 Spectral decomposition of the device output, with $P_{ccl} \sim$ 5 W and $P_{scl} \sim 318 \, \mu$W.

2.15 (b)–(d) the signal is polarized parallel to the pump, while in Fig. 2.15 (a) the signal is both parallel and perpendicular to the pump. The output saturates at higher signal powers because of pump depletion and soliton self-frequency shift effects. The power output at the SCL frequency in Fig. 2.15 (b) (taken without the polarizer) also shows the gain saturation. The small-signal gain and on:off contrast ratio in Figs. 2.15 (c) and 2.15 (d) show the usefulness of the MIPRS as a switch. Gains as high as 10^4 are achieved at low powers, although the gain decreases for increasing signal power because of saturation. On the other hand, the on:off contrast ratio increases with increasing SCL signal power because the net output power increases while the pump leakage at the null is nearly constant.

The large gains are attributed to the wavelength-dependent state of polarization in the fiber. The polarization length $L_p = c/4 \, \Delta n \, \Delta f$ is defined as the distance in which the polarization for two wavelengths become $\frac{1}{2}\pi$ out of phase. In these experiments $L_p \sim 30 \, \mathrm{m}$, which means that the pump and sidebands will be arbitrarily polarized relative to one another after 500 m. Therefore, the quarter-wave plate and the polarizer settings used to null the pump cannot null the fiber output with MI present. Figure 2.16 shows the output spectrum for a pump power of 5 W and signal power of 318 μW, when the angle θ between the input polarization and the principal axes of the fiber is adjusted for maximum output power. Here the gains after the analyzer for the individual components are:

$$G(\omega_0 + 6\pi f) = 28, \qquad G(\omega_0 + 4\pi f) = 95,$$

$$G(\omega_0 + 2\pi f) = 1.4 \times 10^3, \qquad G(\omega_0) = 672,$$

$$G(\omega_0 - 2\pi f) = 476, \qquad G(\omega_0 - 4\pi f) = 56.$$

Since the various wavelengths are cycling through linear and elliptical polarizations every few meters of fiber, the ratio of the various spectral components will change with fiber length. The asymmetry between the upper and lower sidebands is attributable to the soliton self-frequency shift effect.

2.3 Nonlinear directional couplers

A nonlinear directional coupler (NLDC) exhibits sharper switching characteristics than a Kerr gate, although it is typically used

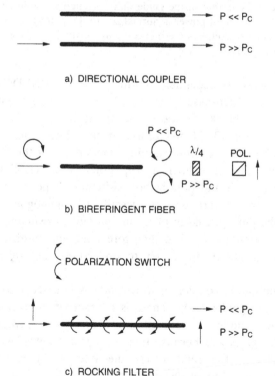

Fig. 2.17 Three analogous implementations of a nonlinear directional coupler. (a) Dual-mode fiber; (b) polarization instability in a birefringent fiber; and (c) a periodic rocking filter in fiber [2.32].

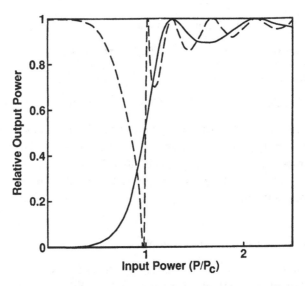

Fig. 2.18 Calculated transmission function for a nonlinear directional coupler at the output of the guide that is excited at the input
(guide #1). The input power is normalized to the critical power P_c.
The length of the couplers are one coupling length L_c (solid line)
and $2L_c$ (dashed line) [2.15].

as a single-input, intensity-dependent routing switch. The NLDC is an
example of a two-coupled-mode system, where the intensity dependent change in index blocks the normal coupling between guides to
cause switching. Jensen [2.15] first proposed and gave a theoretical
treatment of the NLDC. Experimental implementations of the NLDC
(Fig. 2.17) include dual-core fibers [2.16], polarization switching in
birefringent fibers [2.17] and polarization switching in periodic fiber
filters [2.18]. All three experiments are described by analogous equations and show the pulse break-up problem inherent to switching with
an instantaneous nonlinearity. The dual-core fiber and birefringent-
fiber-polarization instability are easier to understand and raise the
major issues.

Consider a dual-mode coupler in which light is coupled into only
one of the two waveguides (#1). There is a coherent interaction of
the two optical waveguides in close proximity, and these waveguides
periodically exchange power because their evanescent fields overlap.
Jensen [2.15] derived the relation for the fraction of power in
waveguide #1 as a function of distance along the guide, z, to be of
the form

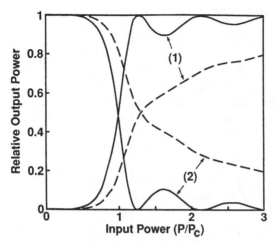

Fig. 2.19 Calculated fractional output power emerging from waveguides (1) and (2) as a function of input power for a coupler of length L_c. Solid curves: constant-intensity input signal. Dashed-dotted curves: coupler response integrated over a $sech^2 t$ pulse intensity profile [2.16].

$$\frac{P_1}{P} = \frac{1}{2}\left\{1 + cn\left[\frac{\pi z}{2L_c}\left|\left(\frac{P}{P_c}\right)^2\right]\right\}, \tag{2.6}$$

where $cn(\phi|m)$ is the Jacobi elliptical function and L_c is the coupling length. The coupling length is a function of the geometry and separation between the two waveguides. The critical power P_c is given by

$$P_c = \frac{\lambda A_{eff}}{n_2 L_c} \tag{2.7}$$

where A_{eff} is the effective area of the waveguide and P_c corresponds to the power needed for a 2π nonlinear phase shift in a coupling length. Note that the critical power is inversely proportional to the coupling length. In the dual-mode coupler, at low intensities the light couples back and forth periodically between the guides with a periodicity of $2L_c$. As the power is increased, the power continues to oscillate between the guides, complete power transfer is still possible, but the period of the oscillation grows with power. As the power increases toward the critical value of P_c, the period approaches infinity, which means that the signal is equally divided between the two waveguides. At higher values of power, the propagation resumes being periodic, but the period of the oscillation decreases with

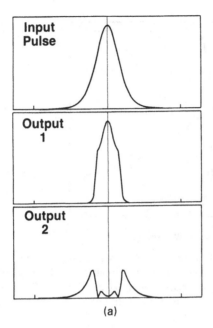

(a)

Fig. 2.20 Calculated reshaping and break-up of $sech^2 t$ pulses due to the nonlinear transmission in Fig. 2.18. (a) Coupler length L_c, peak power of $2P_c$ [2.19]. (b) Coupler length $2L_c$ and peak powers of $0.75P_c$, $0.825P_c$, $0.975P_c$, $1.1P_c$, $2P_c$ and $3P_c$ [2.17].

increasing power and the power transfer between the guides is no longer complete. For example, if the dual-mode coupler is one coupling length long, then a low-intensity signal is coupled across to the other guide while a high-intensity signal remains in the original guide. Figure 2.18 shows the relative output from waveguide #1 as a function of input power for a cw input, and we find a sharp slope and peak transmission as large as one. As the overall length of the coupler increases, the transfer function becomes increasingly sharp around P_c. For example, we include in Fig. 2.18 a dashed curve showing the characteristics for a $2L_c$ long device, which is much sharper than the behavior for one coupling length shown by the solid curve.

Whereas the curves of Fig. 2.18 are plotted for continuous waves, most experiments with ultrashort pulses use Gaussian or squared hyperbolic secant pulses. When the bell-shaped pulses are switched in an NLDC, the weaker wings behave differently from the center peak and the pulse shape breaks up at the output. The result is a reduction

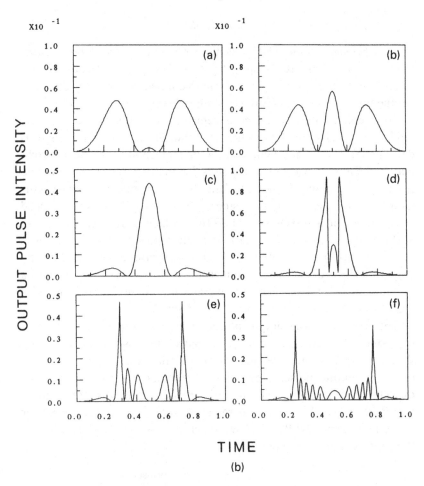

Fig. 2.20 (*continued*)

in the contrast ratio when the output is averaged over the entire pulse. For example, we show in Fig. 2.19 the calculated switching response for pulses with squared hyperbolic secant intensity profiles and a one-coupler-length device, and the degradation of the switching performance is evident. Furthermore, the NLDC also turns out not to be cascadable because the output pulse shape is different from the input. Figure 2.20 shows various temporal profiles calculated at the output of NLDC's that are (a) one- and (b) two-coupling lengths long. Potential solutions of this problem will be addressed at the end of this section.

Two-core fiber directional coupler

Friberg, et al. [2.16], demonstrated a two-core fiber directional coupler using 100 fs optical pulses. Their coupler consists of a
5 mm length of dual-core fiber, which contains two 2.8 μm diameter
germanium-doped cores (core-cladding index difference of 0.003) with
8.4 μm separation between core centers. Each fiber core is single
mode for wavelengths longer than 500 nm, and the coupling length is
determined by a white-light measurement technique to be approximately 4.7 mm at 620 nm. The pulses for the experiment are derived
from a colliding-pulse mode-locked dye laser and a copper-vapor-
laser pumped dye amplifier system (Fig. 2.21). The laser produced
100 fs pulses at a wavelength of 620 nm, which were amplified at an
8.6 kHz repetition rate to 100 nJ. Amplified pulses were focused into
one fiber core, and the other input core was carefully blocked by the
edge of a razor blade. The output from each core was focused onto a
separate power meter, and the average power emerging from each
core was measured as a function of the input power.

The fraction of the output power emerging from each of the two
waveguides is illustrated in Fig. 2.22. Since the power meters respond
slowly compared with the pulse duration, the response is integrated
over the entire pulse. Reasonable agreement is found with the
expected response for sech^2 intensity profiles that was predicted in
Fig. 2.19. From the data, Friberg, et al., estimate the critical power
to be approximately $P_c = 32$ kW, which is relatively high because of
the short length of the coupler.

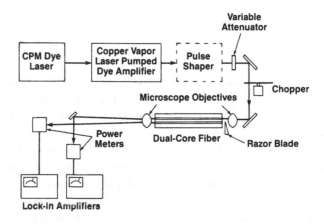

Fig. 2.21 Schematic of experimental apparatus for testing femtosecond switching in dual-core fibers [2.19].

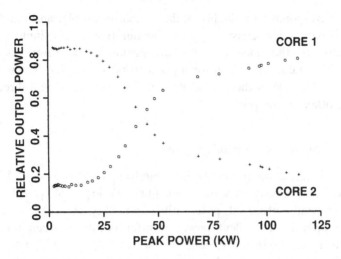

Fig. 2.22 Measured fractional output power from waveguides (1)
and (2) for the 5 mm, dual-core-fiber nonlinear coupler. These data
are the response for 100 fs input pulses [2.16].

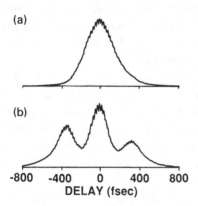

Fig. 2.23 Autocorrelation traces for output pulses from the coupler.
The input pulses were 200 fs in duration. (A) Guide (1), peak power
$P \simeq 2P_c$. (B) Guide (2), peak power $P \simeq 2P_c$ [2.16].

In addition, the reshaping of pulses as seen in Fig. 2.20 (a) is also
confirmed experimentally. For example, Fig. 2.23 illustrates the autocorrelation measurements obtained for 200 fs pulses at a power of
$\simeq 2P_c$ at the output of the nonlinear coupler for (a) the same guide
as the input (#1) and (b) the other guide. While the pulses from
waveguide #1 are similar to those at the input, the pulses from the
other guide are strongly reshaped. The triply peaked autocorrelation

trace corresponds to a doubly peaked intensity profile, and the 340 fs peak separation is consistent with the duration of the input pulse. Furthermore, the individual peaks are significantly narrower than the input. These data show that for input powers greater than P_c the non-linear action selects the central part of the input pulse and directs it to the other output port.

Polarization instability switch

Polarization instability in a birefringent fiber (Fig. 2.17 (b)) behaves analogously to a dual-core fiber. Trillo, et al. [2.17], have experimentally observed polarization instability in a fiber, where strong intensity-dependent power transfer occurs between the two counter-rotating circularly polarized waves in the fiber. For their experiments, the fiber is $2L_c$ long, a left circular analyzer is placed at the fiber output, the input wave is right circularly polarized, and the fractional transmitted power is measured. For this case the predicted reshaping and break-up at the fiber output for a sech^2 input intensity was shown in Fig. 2.20 (b). Because the transfer function illustrated by the dashed curve in Fig. 2.18 is much more complicated than for the one L_c coupler used in the dual-core fiber experiments, the pulse reshaping is also much more complicated.

The polarization instability experiment uses pulses from a fre-quency doubled ($\lambda = 532$ nm) Nd:YAG laser emitting a train of Q-switched, mode-locked pulses (80 ps pulse width at a 80 MHz repetition rate). Light was coupled into and out of the fiber with microscope objectives, and the polarization of the input light was controlled by a linear polarizer followed by a quarter-wave plate. A Babinet–Soleil compensator and a linear analyzer were used at the fiber output to isolate the left circularly polarized wave (Fig. 2.17 (b)). The measurements were performed on a 53 cm step index, aluminum-doped fiber whose index difference was $\delta n = 0.006$ and core diameter was 4.5 μm. The fiber was carefully straightened in a glass tube to minimize the influence of any bend-induced strain or twist, and the length corresponded to about 0.9 times the beat length. The fiber axes do not have to be identified because the experiments use circularly polarized light.

Figure 2.24 shows the experimentally measured reshaped pulse envelope at the fiber output as a function of increasing power. The theoretical envelopes have been computed by averaging over the pulse: since a bell-shaped pulse contains a range of intensities, the

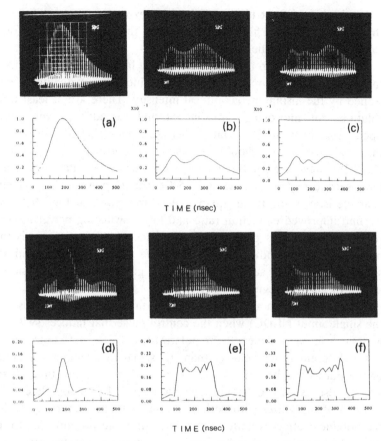

Fig. 2.24 Measured (scope traces) and calculated (solid lines) pulse envelopes from a birefringent fiber exhibiting polarization instability. The different curves correspond to different power levels: (a) linear conditions; (b) peak power P of 300 W (calculation $P = 0.75P_c$); (c) $P = 350$ W (calculation $P = 0.825P_c$); (d) $P = 700$ W (calculation $P = 0.975P_c$); (e) $P = 1.5$ kW (calculation $P = 2P_c$); and (f) $P = 1.9$ kW (calculation $P = 3P_c$) [2.17].

switching is not nearly as dramatic as in Fig. 2.18. At low powers an undistorted fraction of the input envelope is transmitted through the analyzer (Fig. 2.24(a)). As the power is increased, first one dip and then two dips appear in the pulse envelope. For a peak power close to the critical power P_c, the polarization instability leads to a large increase in the corresponding transmitted pulses (Fig. 2.24(d)). At yet higher powers the transmission decreases for the peak of the pulse. In this regime the device acts as a limiter, and the output power is broadened with further increasing power (Figs. 2.24(e),(f)).

Since the birefringent fiber used in these experiments is much longer than in the dual-core fiber example above, the critical power is found to be smaller at a value of $P_c \simeq 800\,\mathrm{W}$.

The pulse break-up problem exemplified by the NLDC is a universal problem for all-optical interactions since the switching is controlled by the instantaneous optical intensity. There are at least three solutions for improving the contrast and cascadability for pulsed operations. First, if the input pulses are square-shaped pulses, then the ideal cw-response should be obtainable. For example, Weiner, et al. [2.19], have used pulse shaping techniques in the picosecond and femtosecond regime to generate $\sim 540\,\mathrm{fs}$ square pulses. The switching characteristics using these square pulses are shown in Fig. 2.25, and we find improved extinction ratio and lower switching power as compared with Gaussian pulses. One drawback of this technique is that pulses of finite bandwidth will still have finite risetimes, and the square edges will be adversely affected by group-velocity dispersion in a fiber or glass waveguide.

A second solution applies to a three-terminal device (as opposed to the single input NLDC) when the control pulse that induces switching is at a different wavelength or polarization. Since the control pulse has a different group-velocity than the signal, the length and timing of the pulses can be adjusted so the control pulse slides through the signal pulse as both travel through the device. For a complete walk-through of the pulses with both pulses maintaining uniform intensity, the nonlinear effect is uniform over the pulse and proportional to the integral of the control pulse intensity. Although this technique circumvents the pulse break-up problem, it does not prevent pulse distortion because of group-velocity dispersion or self-phase modulation. As a general rule of thumb, the slip-through method can be used as long as the device length is shorter than the soliton period Z_0.

A third and perhaps more attractive solution is to use fundamental soliton pulses that have a uniform phase shift, so entire pulses switch as a unit. Now the device must be in the anomalous group-velocity dispersion regime, but the problems of dispersion and self-phase modulation are kept in balance. For example, Trillo, et al. [2.20], have predicted that solitons can improve the NLDC performance when the soliton period is on the order of the coupling length. However, because of the typically short lengths and high switching energies of pulses used with NLDCs, experiments with solitons in NLDCs have not been reported. In fact, although longer coupling lengths would reduce the switching energy, obtaining long L_c is impractical

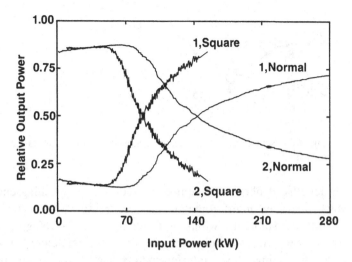

Fig. 2.25 Improved switching in a nonlinear coupler by using square optical pulses. Plots show the relative output power from the two cores (marked 1 and 2) for 100 fs normal bell-shaped input pulses (normal) and for 540 fs square pulses (square) [2.19].

both because the tolerances for fabrication are then too tight and because the device becomes extremely sensitive to external perturbations such as bends. Solitons have been successfully applied to fiber-loop mirrors to avoid the pulse-reshaping problem.

2.4 Mach–Zehnder interferometer switches

Another example of a dual-mode device is a nonlinear Mach–Zehnder interferometer, where the phase difference between the two channels varies with optical intensity. Whereas an NLDC requires a 2π phase shift to switch, an interferometer requires only a π-phase shift difference between the two arms to change from constructive to destructive interference. However, the response of an interferometer follows a squared sinusoid function of the nonlinear phase, which is less sharp than an NLDC. Mach–Zehnder interferometers are used in electro-optic switches such as lithium-niobate devices, where the phase shift required for switching is derived from an applied electric field. In all-optical interferometers, the required nonlinear phase shift will be generated by self- or cross-phase modulation. Let us start by reviewing the single input operation of a nonlinear loop mirror and then move to various configurations for two-input (control and signal) operation.

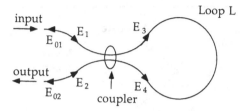

Fig. 2.26 Schematic of the configuration for a nonlinear optical loop mirror [2.22].

Long lengths of fibers are required to reduce the switching energy for Mach–Zehnder interferometers. However, since the interferometer must be stable to within a fraction of a wavelength, it is difficult to use two physically separate fibers to implement a long interferometer. Single-fiber interferometers based on either time-multiplexed signals [2.21] or orthogonally polarized modes of a fiber [2.11] have been proposed and demonstrated. An implementation of the interferometer that has received much attention is based on a nonlinear Sagnac interferometer or a nonlinear optical loop mirror (NOLM). As depicted in Fig. 2.26, the NOLM consists of a four-port directional coupler in which two ports on one side are connected by a loop of fiber. The two arms of the interferometer correspond to the two counter-propagating directions around the loop, and this configuration is very stable since both arms involve exactly the same optical path. When the coupler divides the input equally, the NOLM acts as a perfect mirror. For other than the 50:50 splitting ratio, the nonlinear phase shift is different in the two directions, and the NOLM acts as an intensity dependent mirror. Doran and Wood [2.22] first proposed the NOLM to demonstrate that solitons permit complete pulse switching and that a soliton may act as a fundamental data bit in ultrafast optical processing.

To understand the operation of an NOLM, it helps to review first the governing equations [2.22]. The equations connecting the input and output field shown in Fig. 2.26 for an X coupler are

$$E_3 = \sqrt{\alpha}\, E_1 + i\sqrt{1-\alpha}\, E_2, \qquad E_4 = i\sqrt{1-\alpha}\, E_1 + \sqrt{\alpha}\, E_2, \qquad (2.8)$$

where the power coupling ratio for the X coupler is $\alpha : 1 - \alpha$. Consider the case of a single input at port 1, E_{in}, and neglect any interaction between counter-propagating fields: i.e., the description covers the case when the pulse duration is short compared with the loop length.

After they travel around the loop of length L, the fields E_3 and E_4 are given by

$$E_3 = \sqrt{\alpha}\, E_{\text{in}} \exp(i\alpha|E_{\text{in}}|^2 2\pi n_2 L/\lambda), \qquad (2.9\,(a))$$

$$E_4 = i\sqrt{1-\alpha}\, E_{\text{in}} \exp[i(1-\alpha)|E_{\text{in}}|^2 2\pi n_2 L/\lambda]. \qquad (2.9\,(b))$$

To calculate the outputs E_{o1} and E_{o2} we make the transformations $E_3 = E_4^*$ and $E_4 = E_3^*$ and after some manipulation obtain at the output of port 2:

$$|E_{o2}|^2 = |E_{\text{in}}|^2\{1 - 2\alpha(1-\alpha)(1 + \cos(1-2\alpha)|E_{\text{in}}|^2 2\pi n_2 L\lambda])\}. \qquad (2.10)$$

This equation shows that if α is not one-half then all the power emerges from port 2 whenever

$$\frac{2n_2\pi|E|^2 L}{\lambda} = m\frac{\pi}{1-2\alpha} \qquad (2.11)$$

for m odd, which yields therefore the switching power. The minimum transmitted power occurs for m even and is given by the linear output power, i.e.,

$$|E_{o2}|^2 = |E_{\text{in}}|^2[1 - 4\alpha(1-\alpha)], \qquad (2.12)$$

which determines the contrast ratio. The best switching ratio (contrast between the linear off and the high intensity on) occurs for α closest to $\frac{1}{2}$, but the switching energy increases correspondingly. The transmission function for a cw is illustrated in Fig. 2.27, and we find that the cw transmission coefficient $T_{\text{cw}} = |E_{o2}|^2/|E_{\text{in}}|^2$ has a peak value of 100%.

Although the transmission function for non-soliton and soliton pulses can be calculated exactly using the above relations, Blow, et al. [2.23], have shown approximate formulas that agree remarkably well with exact computer results. They assume that even when the launched pulse is not an exact soliton, they can still use a characteristic phase that applies to the whole pulse. For an input pulse of the form $A \operatorname{sech} t$ the soliton phase is given by

$$\phi_s = \begin{cases} 2(A - \frac{1}{2})^2 & (A > \frac{1}{2}), \\ 0 & (A < \frac{1}{2}), \end{cases} \qquad \begin{matrix} (2.13\,(a)) \\ (2.13\,(b)) \end{matrix}$$

where the standard soliton normalizations have been used (see Appendix A). From relations (2.10) and (2.13(a)) we calculate the

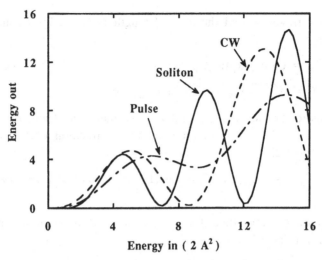

Fig. 2.27 Theoretical transmission functions of a nonlinear optical loop mirror for solitons, non-soliton pulses, and cws or square pulses ($\alpha = 0.42$, $L = 5.8Z_0 = 9.1z_c$) [2.23].

soliton pulse transmission coefficient T_s for an input pulse of amplitude A as

$$T_s = 1 - 2\alpha(1-\alpha)(1 + \cos\{2L[(2\alpha - 1)A^2 - A(\sqrt{\alpha} - \sqrt{1-\alpha})]\}). \tag{2.14}$$

Using the same normalizations, the relation (2.10) can be rewritten for a cw of amplitude A as

$$T_{cw} = 1 - 2\alpha(1-\alpha)\{1 + \cos[L(2\alpha - 1)A^2]\}, \tag{2.15}$$

and the transmission for a hyperbolic secant pulse that is not a soliton, where each portion of the pulse behaves independently, is

$$T_p = \tfrac{1}{2} \int \operatorname{sech}^2 t \, T_{cw}(A \operatorname{sech} t) \, dt. \tag{2.16}$$

Figure 2.27 compares the cw case with both soliton and non-soliton pulses calculated using relations (2.14–2.16) with $\alpha = 0.42$ and a loop length of 5.8 soliton periods. The degradation in contrast and peak transmission is apparent when ordinary pulses are used. The soliton response reaches nearly 100% transmission, which means that the entire pulses switch. The transmission function for cw and solitons are similar, except that the peaks occur at different positions since the soliton uses an average phase over the pulse.

Solitons in optical loop mirrors

Experiments confirming these predictions have been performed by Blow, et al. [2.23], and my colleagues and myself [2.24]. As an example, we observe complete switching of 310 fs soliton pulses at an energy of ~55 pJ with 90% peak transmission in an NOLM. A passively mode-locked NaCl color-center laser [2.7] provides τ = 310 fs Gaussian pulses at λ = 1.692 μm. In the experimental apparatus of Fig. 2.28, a variable attenuator is used to vary the input power, and an isolator prevents feedback into the laser. A 1 cm thick uncoated quartz beam splitter picks off a fraction of the input and reflected light, and apertures are used to block multiple reflections. The NOLM is made of 25 m of single-mode, polarization-maintaining, dispersion-shifted fiber. The coupler is made of quartz blocks in which the polarization-maintaining fiber is glued and polished. Thirty-two percent of the light is coupled over, and the loss across the coupler is ~8%. The loop is closed at a ~90% transmitting splice. Throughout the set-up, the polarization extinction ratio is better than 15:1.

To have enough group-velocity dispersion in the fiber for soliton reshaping and formation, the NOLM must be several soliton periods

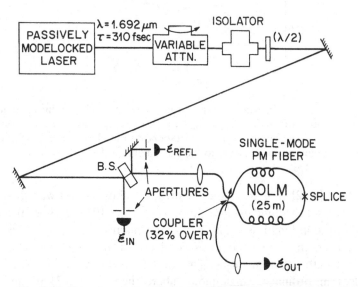

Fig. 2.28 Experimental configuration for the fiber nonlinear optical loop mirror (NOLM) ($\lambda/2$ = half-wave plate, BS = beam splitter).

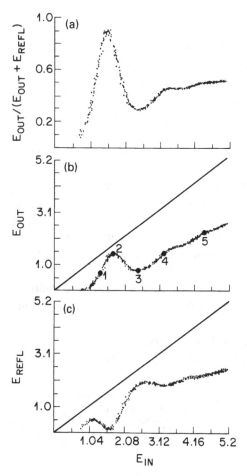

Fig. 2.29 Measured transfer characteristics of the nonlinear optical loop mirror. (a) $E_{out}/(E_{out} + E_{refl})$, (b) E_{out}, and (c) E_{refl} versus E_{in}. The numbers in (b) correspond to the plots in Fig. 2.30. All energies are normalized to the fundamental soliton energy $E_1 = 33.2 \, \text{pJ}$.

Z_0 long [2.22]. For this experiment, the soliton period was $Z_0 = 5 \, \text{m}$ and the soliton fundamental power was $P_1 \cong 107 \, \text{W}$. Thus, the normalized loop length is $\sim 5Z_0$ and a maximum of $\sim 5P_1$ power is coupled from the laser. Sub-picosecond pulses are used to reduce the fiber length requirements ($Z_0 \propto \tau^2$). For example, if we used 10 ps pulses in ordinary, single-mode fiber, then $Z_0 \sim 2.7 \, \text{km}$, and the NOLM would be more than 10 km long. The losses in a fiber of such long lengths would be non-negligible, and the device would be more sensitive to thermal and acoustic fluctuations.

Figure 2.29 illustrates the nonlinear transmission (E_{out}) and reflection (E_{refl}) as a function of input power (E_{in}). The ratio of E_{out} to ($E_{out} + E_{refl}$) increases up to a peak and is then followed by a dip and an approximately flat region. After correcting for the various losses, the peak occurs at $E_{in}/E_1 \sim 1.66$ and the transmission at the peak is $\sim 90\%$. This peak corresponds to a pulse switching energy of $1.66 P_1 \times 310\,fs \cong 55\,pJ$. The peak and null are less pronounced than in the ideal case [2.22] because of soliton self-frequency shift effects [2.12, 2.13]. Since the coupler unequally divides the light in the two directions, the intensity-dependent frequency shift is different in the two directions, and the interference is incomplete. At higher powers, the transmission is nearly flat since the two counter-propagating pulses no longer meet at the coupler because their unequal frequency shifts separate them both temporally and spectrally. In the limit of no-pulse overlap, the expected transmission is $(0.32)^2 + (0.68)^2 = 56\%$. In the experiments by Blow, et al. [2.23], 710 fs pulses (auto-correlation width of 1.1 ps) are used and a 58:42 coupler connects to the two ends of a 100 m fiber. They observe 93% peak transmission at 46 pJ of energy, but because of their broader pulses they do not find frequency shift effects.

At the peak of transmission, we see switching of the complete soliton waveform. In Fig. 2.30, we show the autocorrelations of the transmitted pulses and include inserts of the intensity profile from computer simulations that include the Raman effect at the corresponding power levels. At the peak of the transmission (curve 2), the output is slightly broader than the input pulse but otherwise identical in shape. If the pulses were not solitons, then only the center of the pulse would switch, and the autocorrelation would be narrower. The solitons broaden slightly because, after dividing in the coupler, one arm has lower-than-fundamental soliton power. If the power is lowered further, then the transmitted pulses broaden more (curve 1).

As confirmation of the influence of soliton self-frequency shift, we see pulse break up at higher powers. The frequency shift effect asymmetrizes the pulses, causes higher-order solitons to split, and, as already discussed, temporally separates the pulses from the two directions. The transmitted pulse starts to change shape around the dip in transmission (curve 3). At yet higher powers, the output splits into three pulses (corresponding to five peaks in the autocorrelation) that separate with increasing power. There is qualitative agreement between the calculations and the observed splitting. Note that, because higher-order solitons no longer have uniform phase shift over the

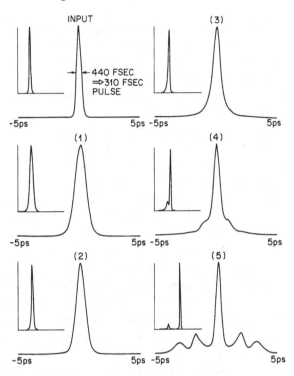

Fig. 2.30 Autocorrelations of the transmitted output for different input powers. The numbers correspond to power levels shown in Fig. 2.29 (b). The inserts are intensity curves versus time from numerical simulations including the Raman effect.

entire pulse [2.25], we do not expect complete switching at higher powers (see Appendix A).

Loop mirror as a three-terminal device

Thus far we have described the NOLM as a single-input device, which can act as a saturable absorber or a pulse shaper. By adding a control beam that is orthogonal either in frequency or polarization, the NOLM can also act as a three-terminal switch. The idea of using a Mach–Zehnder interferometer as an all-optical gate was first discussed by Lattes, et al. [2.26], and their structure was an all-optical analog of electro-optic devices. As shown in Fig. 2.31, in an exclusive-OR (XOR) gate the control pulses propagate through both arms and interfere destructively at the output junction. When either signals *A* or *B* are injected orthogonally polarized to the control, then one arm is phase shifted by π so the two control pulses now interfere

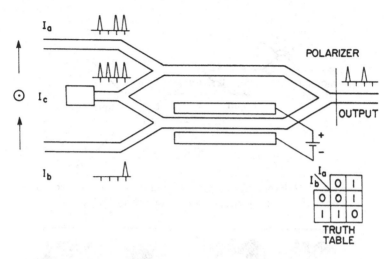

Fig. 2.31 Schematic of a Mach–Zehnder interferometer used as an optical logic gate. A continuous stream *c* of pulses is modulated by the information carrying pulses incident in waveguides *a* and *b* [2.26].

constructively at the output. However, when both pulses are incident, then both arms receive equal phase shifts with no resulting output. A polarizer is used at the output of the interferometer to block both signal inputs. Lattes, et al., first demonstrated the concept in a lithium-niobate device, although the same basic idea can also be applied to a fiber-loop mirror.

The simplest technique for adding a control pulse is to use two separate wavelengths in the NOLM, which is analogous to the operation of the Kerr gates. In this mode of operation, a high-power signal at one wavelength switches a low-power signal at another wavelength, and the device behaves like an all-optical modulator. Blow, et al. [2.27], demonstrated the two-wavelength operation of the NOLM in the configuration of Fig. 2.32. The low-power signal ($\sim 5\,\mathrm{mW}$) was obtained from a cw color-center laser ($1.53\,\mu\mathrm{m}$), and the pump or control was derived from a $1.3\,\mu\mathrm{m}$ mode-locked Nd:YAG laser (pulse width 130 ps). Coupler 1 has a 50:50 splitting ratio for $1.53\,\mu\mathrm{m}$ and provides a monitoring point for the backreflected signal. Coupler 2 combines the $1.3\,\mu\mathrm{m}$ and $1.53\,\mu\mathrm{m}$ signals, and the resultant signal is launched into the NOLM. The NOLM coupler (#3) has a coupling ratio of 50:50 for the $1.53\,\mu\mathrm{m}$ signal and virtually 100:0 for the $1.3\,\mu\mathrm{m}$ control. Therefore, the cross-phase modulation induced phase shift is different in one direction than the other when the control pulse is added. The NOLM uses a 500 m polarization-maintaining fiber, and

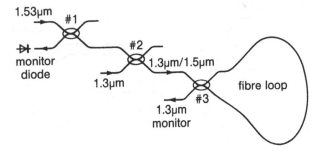

Fig. 2.32 Experimental configuration for testing the two-wavelength operation of a nonlinear optical loop mirror [2.27].

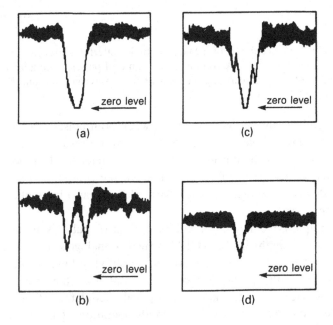

Fig. 2.33 Reflected 1.53 μm signal with a loop length of 500 m for average pump input powers of (a) 20 mW, (b) 35 mW, and (c) 60 mW. In (d) the loop length is 100 m and the average pump input power is 55 mW [2.27].

the response of the device is monitored at coupler 1 with a photo-diode that had a pulse response of 70 ps.

The experimental results for several pump powers are displayed in Fig. 2.33. At an average (peak) power of 20 mW (2 W) at 1.3 μm (Fig. 2.33(a)) there is almost complete switching of the 1.53 μm probe signal. The flat-topped feature seen in the response results from group-velocity walk-off between the two beams, which leads to a uniform phase across the center of the pulse [2.27]. As the pump power is further increased (Figs. 2.33 (b) and (c)), the behavior reflects the periodic intensity response of the NOLM, as we found in Fig. 2.27. In Fig. 2.33 (d), the central part of the pulse is fully reflected, and for this case Blow, et al., use a 100 m NOLM and an input pump power of 55 mW.

Cross-polarized control and signals in a loop mirror

An alternate way of creating a three-terminal switch is to use a control pulse orthogonal to the signal pulses in a polarization-maintaining fiber. Because of the birefringence or the velocity difference between the two axes, the control and signal pulses can be arranged to walk completely through each other. The two pulses interact through cross-phase modulation in the fiber and phase shift one another. Even if the phase shift in a single pass is less than π, the required π-phase shift can be accumulated by using cross-splices (splices in which the slow and fast axes of the fiber are exchanged) to permit multiple passes.

Moores, et al. [2.28], first proposed a multiple slip-through scheme, and their configuration is shown in Fig. 2.34. Their design involves using soliton pulses in which the control pulse slides through half of the signal pulse 11 times until the latter accumulates a π-phase shift. The control pulse always leaves the device through the

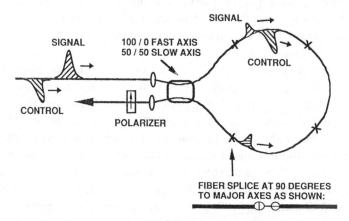

Fig. 2.34 Schematic of a nonlinear optical loop mirror in which orthogonally polarized control and signal pulses interact repetitively because of cross-splices in the fiber [2.28].

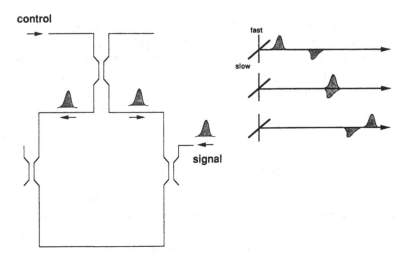

Fig 2.35 Schematic of nonlinear optical loop mirror in which orthogonally polarized control and signal pulses interact and the signal pulses enter and exit through separate polarizing beam splitters [2.29].

transmitted port, but the signal is transmitted only when the control is applied. If a polarizer is placed at the transmitted port along the signal axis, the device acts as an AND-gate. In preliminary experiments Moores, et al., demonstrated some enhancement of the output with the addition of the control, but the contrast ratio was poor because of improper couplers. Nonetheless, a key feature of the configuration is that this switching mechanism allows for regenerative, synchronizing devices, since incoming pulses can be replaced by locally generated pulses.

Avramopoulos, et al. [2.29], point out that another advantage of the slip-through interaction is that solitons or square pulses are not required to obtain complete switching, as we found for the NLDC. Since the nonlinear interaction is proportional to the integrated intensity that lies within a certain time window, which is fixed by the fiber length and the birefringence, the signal pulse may arrive at any time within the window and still have the same effect. Because of this integration effect, an arbitrarily shaped pulse can obtain a uniform phase shift and switch with good contrast ratio. Furthermore, the system can be insensitive to timing jitter in the signal stream if the pulse length is shorter than the window length.

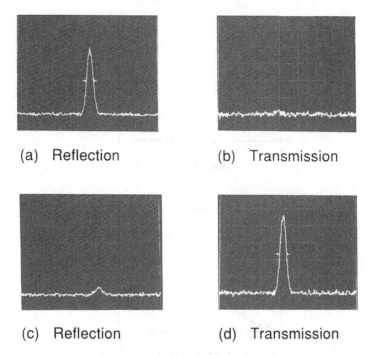

(a) Reflection   (b) Transmission

(c) Reflection   (d) Transmission

Fig. 2.36 The transmitted and reflected outputs from the nonlinear optical loop mirror of Fig. 2.4.10 for signal and control pulse durations of 1 ns. Reflection (a) and transmission (b) at 10 ns relative delay. Reflection (c) and transmission (d) when the control and signal pulses overlap [2.29].

A schematic of their slip-through device is shown in Fig. 2.35, and in their experiments Avramopoulos, et al., use polarization-maintaining components throughout the 500 m loop. The control pulse is split and propagates through both directions in the NOLM, while the signal enters through a polarizing beam splitter, traverses the loop in only one direction, and then is removed by another polarizing beam splitter. By using separate polarizing beam splitters, the couplers can be of conventional design; therefore, they avoid the difficulties that Moores, et al. [2.28], had in making couplers that have different ratios for different polarizations. The signal and control pulses were derived from an electronic pulse generator that produced two independent but synchronized electrical pulses with adjustable delay. These electrical signals drive two laser diodes (one at 1.531 μm and the other at 1.534 μm) to produce ~1 pJ optical pulses that were typically between 0.5 ns and 1 ns in width. The laser diode

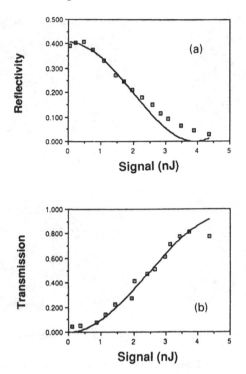

Fig. 2.37 Switching curves showing (a) reflectivity and (b) transmission as a function of signal pulse energy for control and signal pulse durations of 1 ns [2.29].

outputs were then amplified in erbium-doped fibre amplifiers (EDFAs), and to achieve large pulse energies (> 1 nJ) the duty cycle of the lasers were kept low. To overcome the imperfect extinction of the signals in the polarizing beam splitters, a narrow-band interference filter was used to distinguish the signals from the two diodes.

Figure 2.36 shows both the transmitted and reflected outputs of the device in its switched and unswitched states using ~ 1 ns pulses at 2.5 MHz repetition rate. After using the interference filter in front of the detector, large contrasts in transmission (~ 40:1) and reflection (~ 10:1) were observed. In Fig. 2.36 (c) the residual reflected light is mostly amplified spontaneous emission from the EDFA, which is always reflected. The switching energy in this experiment was ~ 4.4 nJ, which corresponds to a peak pulse power of 4.4 W. The maximum reflection was measured to be ~ 41% and the maximum transmission was ~ 84%. Since a larger control pulse is required to switch a smaller signal pulse, the configuration in Fig. 2.35 does not exhibit fan-out. However, Avramopoulos, et al., quote a total gain for the device of ~ 840 because they consider the switch and amplifier together as the device. In addition, in Fig. 2.37 we show the variation of reflectivity and transmission of the NOLM as a function of signal energy. As expected for the Mach–Zehnder interferometer, the response behaves like a squared sinusoid of the induced phase shift.

The advantage of this slip-through interaction interferometer over the two-wavelength NOLM is that the control and signal pulses can be at the same center frequency, so the device may be cascadable. In either scheme a uniform phase shift across the pulse can be obtained so long as the birefringence or frequency difference are adjusted so the two pulses walk completely through each other. However, neither technique guards against pulse shape distortions from self-phase modulation or group-velocity dispersion. As we move toward the shorter pulses that are required for terabit-rate switching, the characteristic length or soliton period Z_0 decreases as the square of pulse width, and any device longer than Z_0 must still use solitons. Furthermore, the walk-off length is proportional to the pulse width, so shorter pulses require higher switching intensity or, in the slip-through example, more segments of crossed-axis fibers. For the slip-through case the walk-off length could also be increased by using moderately birefringent fibers, but then it is difficult to locate and cross the axes of the fiber. In the next chapter, we use soliton dragging and trapping in moderately birefringent fibers to implement logic gates.

2.5 Two-module switch based on Manchester-coded solitons

An ultrafast, all-optical routing switch known as a "2-module" [2.30] uses Manchester-coded soliton pulses in a semiconductor waveguide followed by a fiber. Many routing systems fall under the category of extended generalized-shuffle networks, and Richards [2.31] has shown that a 2-module is a simple and complete building block for such networks. As shown in Fig. 2.38(a), a 2-module gate has two inputs and two outputs: the 2-module ORs the two inputs and passes the results to the two output ports only if the gate is enabled by a control pulse. In addition to gating the input, a 2-module switch provides gain and logic-level restoration. By proper design and dilation of the network, we can guarantee that only one input is active at any given time. Therefore, the input section can be implemented using a 3 dB coupler or any variety of wired-ORs. As an example, a 2-module can be implemented using a single symmetric

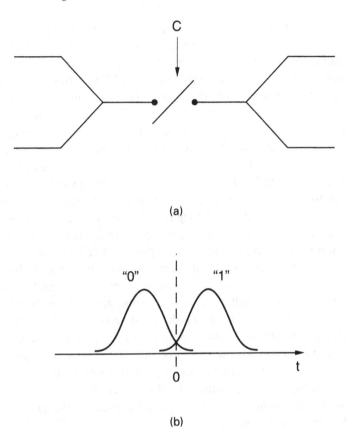

Fig. 2.38 (a) Schematic of 2-module routing gate, which has two in-puts and two outputs. By proper design of the network, we can guarantee that only one input is active at any given time. (b) Manchester-coded soliton pulses in which a "0" corresponds to a declining slope of the pulse at the reference time and a "1" corres-ponds to a rising slope at the reference time.

self-electro-optic effect device, but then the speeds are limited to a few hundred kilobits per second [2.30].

Manchester coding is often used in local area networks to guarantee synchronization and to avoid a direct current component in electronic systems. In the Manchester code, there is a transition at the middle of each bit period, which serves as a clock and also as data: a low-to-high transition represents a "1", and a high-to-low transition represents a "0". We can define analogously Manchester coding of soliton pulses (Fig. 2.38(b)), where a "0" corresponds to a declining slope and a "1" corresponds to a rising slope of the pulse at the reference time. Manchester-coded solitons are similar to soliton-dragging logic that is based on time-shift keying (see Section 3.1). In time-shift keying, a "1" corresponds to a pulse within a specified clock window, and we shift the pulse in or out of the clock window by using soliton dragging. For a Manchester-coded system, the soliton pulse is only shifted to one side or the other of the bit window. In high-speed systems there are two separate issues of clock recovery and sensitivity to timing jitter. The clock recovery is simpler for Manchester-coded pulses since each bit period contains a pulse and a transition. However, the timing constraints may be tighter than in time-shift keying because each pulse must be properly positioned.

We implement the 2-module using a semiconductor waveguide followed by polarization-maintaining fiber. The experimental configuration is the same as a hybrid time-domain chirp switch, which will be described in further detail in Sections 3.2 and 4.2. We place the signal input along one polarization axis, the control pulse along the orthogonal axis, and the polarizer at the fiber output aligned with the control axis. The pulses interact through the nonlinear index in the semiconductor, and cross-phase modulation with no walk-off leads to an attractive force between nearly coincident, orthogonally polarized pulses (further described in Section 4.2). Therefore, the control pulse is incident at the reference time ($t = 0$ in Fig. 2.38(b)), and it shifts in the direction of the signal pulse. There is an output only when the control is applied, and the control copies the value of the signal input; so, this 2-module behaves like a polarization-rotation switch with fan-out.

Physically, when the control pulses "sees" an intensity-dependent index slope across it, the control chirps or shifts its center frequency and, through the group-velocity dispersion in the fiber, the chirp translates into a timing shift. For a positive nonlinear index and anomalous dispersion, the control pulse shifts to regions with higher index, and hence there is an attractive force between pulses. A simple analogy may be a billiard ball (the control pulse) resting on a taunt, level, rubber surface. If there are no inputs, then the ball remains stationary. Adding a signal creates a well or depression on one side of the ball, and the ball tends to roll in that direction.

We tailor the experimental parameters to yield a cascadable gate with a fan-out of at least two. We adjust the control energy to be a fundamental soliton in the fiber; i.e., for a control pulse of 22.6 pJ in the fiber, and we find that the output pulse width equals the input pulse width. However, because of mode mismatch and poor coupling

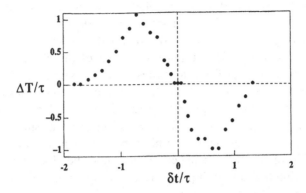

Fig. 2.39 Shift of the control pulse versus initial separation between signal and control pulses ($\delta t = t_{\text{control}} - t_{\text{signal}}$). The control pulse is 56.4 pJ and the signal is 10.5 pJ in the semiconductor waveguide.

between the semiconductor and fiber, the control energy in the semiconductor must be 56.4 pJ to yield a soliton in the fiber. The signal energy is determined by observing the shift of the control pulse as a function of the separation between control and signal pulses. In particular, suppose that a "0" is a signal that arrives about one pulse width early and a "1" is a signal that arrives about one pulse width late. Then, the signal energy should be at least large enough to shift the control by one pulse width. For example, in Fig. 2.39 we plot the normalized shift of the control pulse ($\Delta T/\tau$) versus the initial separation between pulses $\delta t = t_{\text{control}} - t_{\text{signal}}$ for a signal energy of 10.5 pJ in the semiconductor waveguide. In this example the fan-out, which is defined as the control energy out of the fiber divided by the signal energy in the semiconductor, is 2.15.

The cross-correlation of the control output with the clock or reference pulse is shown in Fig. 2.40. In the absence of a signal pulse or if the signal is coincident with the control, then the control arrives at its original position (dotted curve in Fig. 2.40). Note that this does not lead to an error since the next gate will also not shift the control. A "0"-signal input shifts the control pulse to earlier times, while a "1" input shifts the control the other way. Recall that the cross-correlation is wider than the actual signal, and there is even further broadening due to amplitude and timing jitter. Therefore, a one-pulse-width shift does not separate the cross-correlation traces.

Instead of using a control pulse that starts at the reference time, we could also operate the switch with both signal levels equal to the

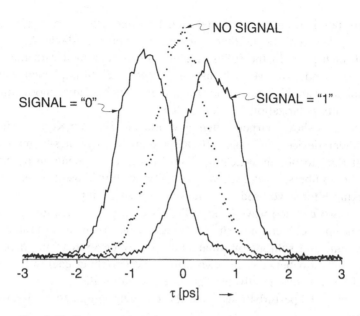

Fig. 2.40 Cross-correlation of control output from the fiber with the clock or reference pulse. The dotted curve in the center is with the signal blocked, the curve to the left is with the signal equal to "0" and the curve to the right is with the signal equal to "1".

control level and both signals Manchester encoded. Then, for any single input the switch simply passes the value to the output along the same axis. On the other hand, if both inputs are active, then the bit streams exchange value; i.e., equal valued signals do not change state while inputs with different values pull each other in opposite directions, thereby exchanging their values. Notice that the interaction is conservative in that the amount by which one pulse increases its photon energy is equal and opposite to the amount by which the other's is decreased. Conservative gates that use solitons will be discussed in more detail in Section 3.3.

2.6 Summary

Several kinds of routing switches use the instantaneous change in refractive index n_2 to cause intensity-dependent switching. In Kerr gates, the nonlinearity is used to cause a change in state of polarization, and Kerr shutters/demultiplexers and optical limiters or fast saturable absorbers are based on this effect. When phase matching is included between two closely spaced frequencies that correspond to the control and signal wavelengths in a single-mode fiber, 4WM can be used to implement a non-cascadable AND-gate with high gain. In the soliton regime, 4WM combined with the non-linearity and group-velocity dispersion were used to implement modulational instability fiber interferometer switches and modulational instability polarization rotation switches.

NLDCs show sharper switching characteristics than Kerr or Mach–Zehnder devices, although NLDCs require a two-times-larger phase shift (2π) to obtain switching. Thus far the experiments using either dual-core-fibers, polarization instability in birefringent fibers or a rocking filter have used the intensity of a single input to detune between two coupled waveguides. NLDCs clearly demonstrate the pulse break-up problem for $sech^2$ or Gaussian pulses in an instantaneous medium, and solitons or square pulses may be used to improve the contrast ratio between the two arms. The switching energy for most NLDCs remains relatively high because fabrication tolerances and sensitivity to perturbations force the coupling length to be relatively short.

Mach–Zehnder interferometers rely on the nonlinear phase change to cause a change from constructive to destructive interference. NOLM's are stable, all-fiber interferometers and can have lower switching energy since long fiber lengths can be used. As a single input device the NOLM acts as a pulse shaper or a fast saturable absorber. Also, the NOLM has been used to demonstrate that solitons switch as a unit, thereby avoiding the previously mentioned pulse break-up problem. By adding a control pulse at a different frequency or a different polarization, the NOLM can also be used as a three-terminal routing or logic module. If the control and signal pulses walk through each other completely, then the phase shift is integrated over the interaction length, and a uniform phase shift can be obtained without using solitons or square pulses. The fiber Mach–Zehnder interferometer implementations described here do not exhibit small-signal gain between the control and signal pulses, although net device gain can be obtained by adding an erbium-doped fiber amplifier. In the next section I will describe cascadable logic gates in which the physical mechanisms lead to small-signal gain without adding an amplifier.

An ultrafast, guided-wave, 2-module switch uses temporal shifts to accomplish routing functions. We find that an attractive potential between soliton pulses can be used to copy the value of the signal onto an orthogonally polarized control beam. When the control beam is not applied, the output is simply disconnected. Since the signal photons are replaced by new control pulses from the laser power supply, the 2-module shows fan-out and logic-level restoration. The gate is also cascadable because the control pulse is a fundamental soliton in the fiber with the output pulse width equal to the input pulse width.

3

Digital soliton logic gates

Whereas the previous chapter concentrated on routing switches, this chapter focuses on all-optical soliton interactions in fibers to implement ultrafast digital logic gates. Three kinds of soliton gates are discussed and compared: soliton-dragging logic gates (SDLG), soliton-interaction gates (SIG) and soliton-trapping logic gates. All these fiber gates have the potential of operating at speeds over 0.2 Tbit/s and have switching energies ranging from 1 to 50 pJ. The soliton-dragging and -trapping logic gates are based on the interaction between two orthogonally polarized pulses in a birefringent fiber, and the mathematical details are included in Appendix B. In particular, soliton-dragging logic gates satisfy all requirements for a digital optical processor and have switching energies approaching 1 pJ. In addition, soliton-dragging logic gates are one example of a more general time-domain chirp switch architecture (TDCS) in which a dispersive delay line acts as a "lever-arm" to reduce the switching energy. Soliton-interaction gates (mathematical treatment in Appendix C) are based on elastic collisions between solitons and illustrate that solitons can be used to implement billiard-ball logic operations. Soliton-trapping logic gates are sensitive to the timing of the input pulses and display on:off contrast ratios as large as 22:1. The soliton-trapping AND-gate (STAG) can serve as the final stage in an all-optical system and as the interface to electronics.

There is a growing trend toward "digital" systems because of their high reliability and fidelity, which is in part due to wide margins for "0" and "1" states. Miller [3.1] and Keyes [3.2, 3.3] have described requirements for devices used in digital systems. The devices must be cascadable (the output of one device must serve as the input to an identical device), must have fan-out (the output of one gate must switch at least two similar gates), and must be Boolean complete (e.g. NOR, NAND gates). In addition, there are desirable attributes that make it convenient to use the gates in systems. Most systems use three-terminal devices, which have three ports that must be orthogonal either in space, time, polarization or wavelength. Logic devices should also have input/output isolation so that a reflection from the next device does not disturb the operation. In clocked systems, the isolation can often just be temporal: i.e., a reflection at the incorrect time has no affect. Finally, and perhaps most importantly, digital devices should have logic-level restoration. Although a range of values may be accepted as a "1" or "0" at the input, the output is a standard "1" or "0". Logic-level restoration prevents accumulation of noise at each stage, making an infinitely extendible system possible.

A brief summary of this chapter follows. The next three sections describe gates based on time-shift keying by using elastic or inelastic collisions between solitons. In Section 3.1 the SDLGs use inelastic collisions between orthogonally polarized pulses, and a more generalized description of the switch is given in Section 3.2 on the TDCS architecture. The SIGs in Section 3.3 are based on elastic collisions between solitons along the same fiber axis, and the SIGs demonstrate a unique phase-independent collision property of solitons after the pulses separate. Then, in Section 3.4 soliton-trapping logic gates provide temporal discrimination and energy contrast in addition to illustrating how solitons can compensate for the polarization dispersion in fibers. Finally, in Section 3.5 we cascade a SDLG and a STAG to generate an energy-contrast output without use of amplifiers or cross-correlators.

3.1 Soliton-dragging logic gates

A low-energy, all-optical, NOR-gate that is based on timing shifts from soliton dragging is demonstrated in moderately birefringent optical fibers [3.4]. Figure 3.1(a) shows a schematic of the NOR-gate that consists of two lengths of fibers (e.g., the first fiber could be $\sim 5Z_0$ and the second between 25 and $35Z_0$, where Z_0 is the soliton period). The two fibers are connected through a polarizing beam splitter, and the output is filtered by a polarizer or another polarizing beam splitter. The power supply or control pulse C provides gain, propagates along one principal axis in both fibers and corresponds to A NOR B at the output. For a cascadable gate, C should be approximately a fundamental soliton and should experience

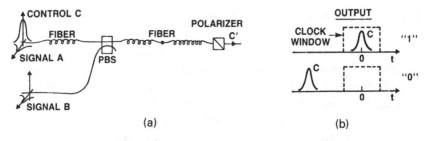

Fig. 3.1 (a) Schematic of a soliton dragging NOR-gate with control or power supply C along one axis and signals A and B polarized orthogonally (PBS = polarizing beam splitter). (b) Example of time-shift keyed logic, where a Boolean "1" corresponds to a control pulse that arrives within the clock window.

insignificant frequency shift from the self-Raman amplification effects [3.5]. The two signal pulses A and B are polarized orthogonal to C and are blocked by the polarizer at the output. The signals are timed so that A and C coincide at the input to the first fiber and B and C coincide (in the absence of A) at the input to the second fiber. Note that logic-level restoration is obtained by replacing the signal pulses by the control pulse at the output. Also, in a system consisting of such gates, the output from one gate (the control pulse) is flipped in polarization and serves as the input to the next gate, and at the output of the second gate is removed from the system by the polarizer. Consequently, the output of one gate passes through only one more level of logic before it is removed from the system.

The logic gate is designed for digital optical applications and operates based on time shifts from soliton dragging. In soliton dragging, two pulses that are coincident in time interact through cross-phase modulation [3.6, 3.7] and, consequently, chirp in frequency and time shift by propagating in a dispersive delay line (cf. Section 3.2). The pulse along the slow axis speeds up, while the pulse along the fast axis slows down. We assume that a signal corresponds to a pulse with guard bands surrounding its time slot. Then, a "1" corresponds to a pulse that arrives within the clock window and a "0" either to no pulse or an improperly timed pulse (Fig. 3.1(b)). For the NOR-gate, the fiber length is trimmed so that in the absence of any signal the control C arrives within the clock window and corresponds to a "1". When either or both signals are incident, they interact with the control pulse through soliton dragging and pull C out of the clock time window.

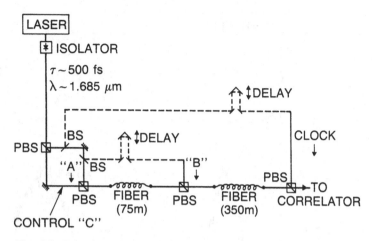

Fig. 3.2 Experimental configuration for testing the all-optical soliton-dragging NOR-gate (BS = beam splitter, PBS = polarizing beam splitter, POL = polarizer).

Figure 3.2 shows the experimental apparatus for testing a single NOR-gate. We obtain $\tau \sim 500$ fs pulses near $1.685\,\mu$m from a passively mode-locked NaCl color-center laser in which a 2 mm thick quartz birefringent plate deliberately limits the bandwidth and, thus, broadens the pulses [3.8]. The input stage separates the control C, signals A and B, and clock beams, and stepper-motor controlled delay stages are used to time properly signal B and the clock. The two fibers are 75 m and 350 m long, have a polarization dispersion of about 80 ps/km, and when carefully handled exhibit a polarization-extinction ratio better than 14:1. The control pulse output and the clock are directed to a correlator to measure the time shifts.

The correlation of the clock with the NOR-gate output is illustrated in Fig. 3.3. The dotted box corresponds to the clock window, and we see that C arrives within this window when no signal is present. When $A = 1$ or $B = 1$, C shifts between 2 and 3 ps out of the clock window; the shift from A is larger since C can time shift in both fibers. When $A = B = 1$, C shifts by about 4 ps. The additional noise and broadening for $A = B = 1$ occurs because A and B, which are parallel polarized, interfere at the polarizing beam splitter; this interference can be avoided by using two separate beam splitters to remove A and introduce B. In this example the fan-out or gain (= control out/signal in) is six and the signal energies are 5.8 pJ each. The control pulse energy in the first fiber is 54 pJ and is reduced to 35 pJ in the second fiber because of coupling losses.

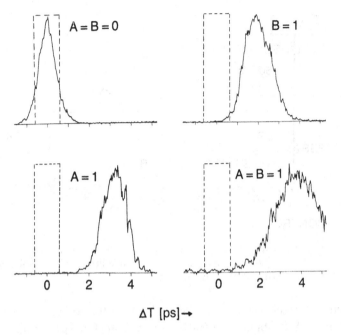

Fig. 3.3 Cross-correlation of clock with soliton-dragging NOR-Gate output. The signal energy is 5.8 pJ and the gain (= control out/signal in) is six.

All-optical multivibrator or ring oscillator

To prove the cascadability and fan-out of the logic gate, we connected the NOR-gate as an inverter and fed the output back to the input (*A* = 0, *B* = previous output from gate). We placed a 50:50 beam splitter at the output and sent half of the output through an adjustable, free-space, delay line to the *B* input. The correlator was set to the center of the clock time window. As Fig. 3.4 shows, with the feedback blocked the output is a string of 1s. When the feedback is added, the output becomes an alternating train of 1s and 0s whose period is twice the fiber latency (1.75 μs). Although the demonstration of a sub-MHz oscillator may not seem terribly impressive, this is an absolute test of both the fan-out and cascadability. After all, what could be a more stringent test of the cascadability that to force a gate to drive itself. Furthermore, in our experiment the device must have at least a fan-out of four since we use a 50:50 beam splitter at the output of the gate and there is an additional 3 dB loss when coupling back into the device because of our bulk optics. Thus, we demonstrate an all-optical multivibrator or ring oscillator using only a logic gate and no amplifiers.

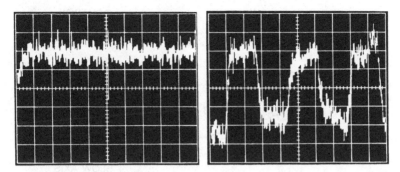

Fig. 3.4 Output from an all-optical multivibrator, which is simply a
NOR-gate configured as an inverter whose output is fed back to the
input. Feedback blocked on the left and feedback added on the
right. The horizontal time scale is 1 μs/div.

Having demonstrated the NOR-gate operation, we now examine
the design criteria for the parameters chosen in our experiment,
which will be further explained by the time-domain chirp switch
architecture of Section 3.2. The maximum control C power is limited
by soliton self-frequency shift (SSFS) or Raman amplification [3.5] if
we want the switch to be cascadable (i.e., to obtain same input and
output frequencies). We adjust C and the fiber length so that the shift
of the pulse center frequency is less than one-sixth of the pulse spectral width. Furthermore, we verify that C is approximately a fundamental soliton by checking that the output pulse width is comparable
to the input. The fiber birefringence should be such that the walk-off
length for orthogonally polarized pulses is greater than or comparable
to a soliton period Z_0 [3.9]; otherwise, the interaction length is too
short, resulting in insufficient soliton dragging. The first fiber must be
more than two walk-off lengths long, while the length of the second
fiber must yield at least a π phase shift from cross-phase modulation,
i.e., $\frac{2}{3}(2\pi/\lambda)n_2 I_{\text{signal}}L \sim \pi$.

For a given control amplitude and fiber length, we study the signal
amplitude range consistent with the system constraints. For instance,
in Fig. 3.5 we plot the normalized shift of the control pulse as a function of signal energy for a control energy of 48 pJ. As long as Raman
amplification effects are minimized, the control pulse can propagate
along either the slow or fast axis of the fiber. A positive delay means
that C arrives later, while a negative delay means that C arrives earlier. The desired width of the clock window sets the minimum signal
energy and maximum gain. In addition, the time-guard band between
pulses sets an upper limit on the signal energy. For example, if we

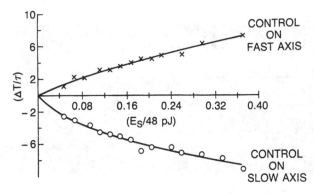

Fig. 3.5 Time shift of the control pulse from soliton-dragging as a function of signal energy. The control pulse energy is 48 pJ and the fiber length is 350 m.

assume a ten-pulse width separation between adjacent signals, which means a maximum bit rate of 0.2 Tbit/s for 500 fs pulses, then the shift should be less than five pulse widths.

3.2 Time-domain chirp switch architecture

A novel time-domain chirp switch (TDCS) architecture [3.10] is a generalization of fiber soliton-dragging logic gates and in which digital logic is based on time-shift keying. By using solitons, with their particle-like behavior, we can separate the frequency change due to cross-phase modulation from the phase and temporal change required for switching. Understanding the TDCS architecture permits us to: (a) optimize individually the frequency change and the phase or temporal change; (b) apply our knowledge to materials other than fibers such as semiconductors and organics; and (c) lower the switching energy to levels approaching a picojoule.

As shown in Fig. 3.6 (a), the TDCS consists of a nonlinear chirper followed by a dispersive delay line and has two orthogonally polarized inputs (signal and control pulses). In the absence of a signal pulse, the control pulse propagates through both sections and arrives at the output within the clock window. For a cascadable switch, the self-induced chirps on the control in both sections must balance, and the output pulse must resemble the input. Adding the signal pulse creates a time-varying index change that chirps the control pulse and shifts its center frequency [3.6, 3.7]. As the chirped control pulse propagates through the soliton-dispersive delay line, the frequency shift is translated into a time change.

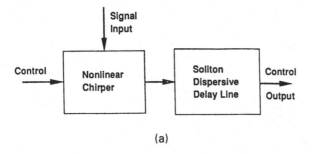

(a)

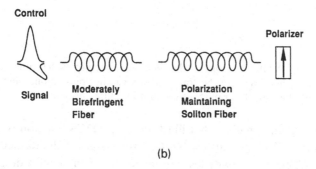

(b)

Fig. 3.6 (a) General architecture for an all-optical time-domain chirp switch (TDCS). The signal creates a time-varying index change in the nonlinear chirper, and the soliton dispersive delay line translates the frequency shift into a time shift. (b) Soliton-dragging logic element is an all-fiber example of a TDCS.

The particle nature of solitons is crucial for separating the TDCS in two sections and relaxing the requirements on the nonlinear material. If the second section were just a linear dispersive delay line (e.g., a pair of prisms or gratings), then changing a "1" to a "0" would require a frequency shift on the order of the entire spectral width. It can be shown that this is the same as requiring the interaction to result in a π-phase shift within the nonlinear chirper [3.10]. Since a fundamental soliton acts as a particle, even a slight shift in the center frequency can cause the complete soliton to shift in time, which means that much less than a π-phase shift results after the interaction between the two pulses. As in most switching configurations, we still need to obtain at least a π-phase shift to change state, but the phase shift is accumulated as the frequency shifted pulse propagates in the dispersive delay line. In other words, although all the interaction occurs in the first section, the phase shift required for switching gathers in the second section.

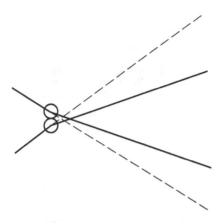

Fig. 3.7 Spatial analogy of the time-domain chirp switch function. The dashed lines are the trajectories for either pulse alone, and the solid lines are the trajectories with both pulses present.

A simple spatial analogy that illustrates the TDCS function is illustrated in Fig. 3.7. Each particle by itself travels along the dashed trajectories. When both particles are incident, then soliton-dragging attracts the two particles and changes their angle of propagation. The change in angle occurs within the "nonlinear chirper," and just after the interaction there is insufficient shift in either pulse for switching. However, as the particles propagate through the dispersive delay line, the change in angle magnifies into a larger separation from the dashed trajectories. Therefore, the dispersive delay line acts as a "lever arm" to translate a small angle change (frequency shift in the SDLG) into a large separation from the original trajectory (time shift in the SDLG). The longer the lever arm, the smaller the change in angle through the interaction needs to be. This is why the TDCS architecture can result in low switching energies at the expense of increased latency or length of the device.

Experimental verification of switch architecture

In a SDLG the nonlinear chirper corresponds to a moderately birefringent fiber, and the soliton-dispersive delay line corresponds to a long polarization-maintaining fiber (Fig. 3.6(b)). We test the TDCS concept in the experimental apparatus of Fig. 3.8, which corresponds to an inverter (the NOR-gate can be considered as a cascade of two inverters). The input stage separates the control and signal pulses ($\tau \sim 500\,\text{fs}$ and $\lambda \sim 1.685\,\mu\text{m}$), permits them to be

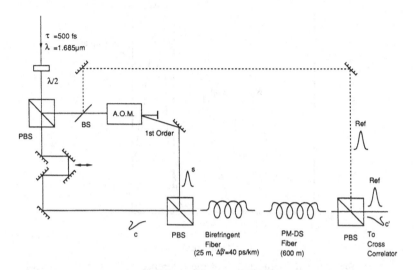

Fig. 3.8 Experimental apparatus for testing a soliton-dragging logic element, where the nonlinear chirper and dispersive delay line are separated in two fibers (PBS = polarizing beam splitter; BS = beam splitter; AOM = acousto-optic modulator; PM-DS = polarization maintaining, dispersion shifted).

independently varied, and then recombines them overlapped in time. Since the signal corresponds to the first-order diffracted beam from an acousto-optic modulator, it interacts with the acoustic wave and has a random phase relative to the control pulse. Therefore, we insure that the switching is not critically dependent on the phase between the control and signal. The two pulses are coupled into a moderately birefringent fiber with a polarization dispersion of $\Delta\beta' = 40$ ps/km ($\Delta n \simeq 1.2 \times 10^{-5}$), and the 25 m length corresponds to about two walk-off lengths. After the interaction, the pulses propagate in a 600 m length of polarization-maintaining, dispersion-shifted fiber that acts as the soliton-dispersive line (the high birefringence in this section prevents interaction between the pulses). Finally, the output control pulse is combined with a reference pulse, and the two are sent to a cross-correlator. Note that unlike the SDLG described earlier, the moderate birefringence is restricted to the first few meters of fiber, which is of practical importance since maintaining polarization extinction in long lengths of moderately birefringent fiber is difficult.

Time-shift keying logic requires the control pulse to be shifted by several pulse widths when the signal is present; although the width of the time window has to be determined by various system constraints,

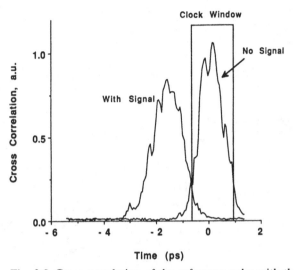

Fig. 3.9 Cross-correlation of the reference pulse with the control output from the soliton-dragging logic element. The box corresponds to the clock window (control energy = 49 pJ, signal energy = 5.6 pJ).

we insist here on a three-pulse-width shift. For example, in Fig. 3.9 we show the cross-correlation between the control and reference pulses for a control energy of 49 pJ and a signal of 5.6 pJ in the first fiber (this corresponds to a peak power of 98 W for the control and 11.2 W for the signal). By itself, the control pulse arrives within the clock window, but adding the signal shifts the control pulse by ~ 3 pulse widths out of the clock window. The phase shift of the control pulse through cross-phase modulation in the first fiber is approximately $\Delta\phi_1 \sim \frac{2}{3}(2\pi/\lambda)n_2 I_s l_{wo}$, where l_{wo} is the walk-off length (~ 12.5 m in our experiments). Plugging in numbers, we find $\Delta\phi_1 \sim \frac{1}{26}\pi$, which is indeed a small fraction of π. Incidentally, if we replace the walk-off length in the previous equation by the second fiber length l_2, we find that the resulting phase shift is $\Delta\phi \sim 1.85\pi$. Therefore, the TDCS may be thought of as a π-phase shift switch except that the effective interaction length is l_2 rather than l_{wo}.

We can also estimate the phase and frequency shift in the first fiber from the measured time shift. The time shift ΔT normalized to the pulse width τ in the soliton dispersive delay line is given by

$$\frac{\Delta T}{\tau} = \frac{D_2 l_2 |\Delta\lambda|}{\tau} = \frac{\lambda^2}{2\pi c\tau} D_2 l_2 |\Delta\omega|, \tag{3.1}$$

where D_2 is the group-velocity dispersion of the second fiber and $\Delta\omega$ ($\Delta\lambda$) is the shift of the center frequency (wavelength) due to the interaction. The maximum frequency shift is related to the phase shift approximately by

$$|\Delta\omega| \sim \frac{\partial\phi}{\partial t} \sim \left|\frac{\Delta\phi_1}{\tau}\right|. \qquad (3.2)$$

Combining the above two relations we obtain the phase shift in terms of the time shift

$$\frac{\Delta\phi_1}{\pi} \simeq \frac{\Delta T}{\tau}\frac{2c\tau^2}{D_2 l_2 \lambda^2}. \qquad (3.3)$$

For parameters in these experiments ($D_2 = 5.65\,\mathrm{ps/(nm\,km)}$ and $\Delta T/\tau \simeq 3$) we obtain $\Delta\phi_1 \simeq 0.0467\pi \simeq \frac{1}{22}\pi$, which is close to the phase shift we estimated above directly from cross-phase modulation. The corresponding frequency shift for these experiments is $\delta\nu_1 \sim 0.0467\,\mathrm{THz}$. Since the measured full width at half-maximum spectral width for the pulses is $\Delta\nu_0 \sim 1\,\mathrm{THz}$, we find $\delta\nu_1/\Delta\nu_0 \sim \Delta\phi_1/\pi$. This reiterates the earlier claim that if the second section of the TDCS were just a linear dispersive line, then separating the pulses would require a spectral width shift, which would return to the $\Delta\phi_1 \sim \pi$ requirement.

Inverter with a picojoule switching energy

A key feature of the TDCS is the low switching energy for short pulse applications. For this two-fiber soliton-dragging case we find that the switching energy E_s is proportional to $(\Delta T/\tau)\tau^2 \Delta n_1/L_2 D_2$, where Δn_1 is the birefringence in the first fiber, and L_2 and D_2 are the length and group-velocity dispersion in the second fiber [3.11]. Thus, the birefringence of the first fiber and the dispersion in the second fiber are the crucial fiber parameters. The switching energy decreases for short pulses because of the $E_s \propto \tau^2$ dependence. One power of τ arises since $E \propto I\tau$, and the second power arises since $\Delta T/\tau$ is a constant (i.e., for shorter pulses less of a physical time shift is required). If $\Delta T/\tau$ is a constant, two ways of reducing E_s are to lower Δn_1 and increase the length (or number of soliton periods) of the second fiber. We cannot reduce the pulse width below $\sim 500\,\mathrm{fs}$ because of soliton self-frequency shift effects [3.5], and we hesitate to increase the dispersion in the second fiber since this increases the soliton energy and, consequently, the control energy.

As a proof of principle, we have demonstrated an inverter with a switching energy as low as 1 pJ [3.11]. For the 1 pJ inverter we use 30 m of moderately birefringent fiber followed by a 2 km length of polarization-maintaining fiber. The birefringence in the first fiber is $\Delta n_1 \sim 10^{-5}$, and the polarization extinction ratio is $\sim 30{:}1$. Therefore, we have lowered the birefringence by two and a half times and increased the fiber length more than threefold. A polarization-maintaining fiber is required in the second section to avoid nonlinear polarization-rotation effects [3.12]. Furthermore, the core size and, consequently, the soliton energy of the second fiber is larger than that of the first fiber.

The correlation of the control output with the reference pulse is shown in Fig. 3.10 (a). The rectangle outlines the clock window, and we see that adding the signal shifts the control pulse out of this window. The signal energy in the first fiber is 1 pJ and the control energy out of the second fiber is 28 pJ, which corresponds to a fan-out of 28. The signal energy is determined by requiring sufficient chirp to shift the control pulse, while the control energy is determined by requiring the control to be a fundamental soliton in the second fiber. Although a 2 km long logic gate may not be desirable, our purpose is to verify that even in fused silica fibers an optical gate based on the nonlinear index can fundamentally approach the 1 pJ switching energy level, which then raises the feasibility of potential applications.

Despite the shift in peak by ~ 4 ps, the contrast within the clock window is limited by the broadening of the cross-correlation. The cross-correlation broadens because of jitter during the 9.7 μs latency through the 2 km fiber and because the control energy in the second fiber is lower than an $N = 1$ soliton. For example, we estimate the control power to be about 0.45 times the $N = 1$ soliton power P_1; while this is still a fundamental soliton, the width adjusts to compensate for the lower power. From the autocorrelation of the control output pulse in Fig. 3.10 (b) we find that the pulse broadens to 1.3 ps (autocorrelation width is $\sqrt{2}$ times the pulse width), which is 2.3 times the input pulse width.

We purposely stay below the $N = 1$ soliton energy in the second fiber to avoid soliton self-frequency shift effects. Frequency shifting deleteriously affects the switch because it translates amplitude jitter into timing jitter and because the gate is not cascadable if the

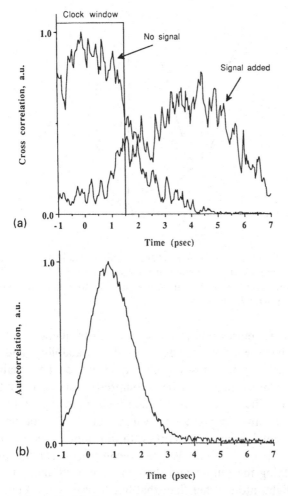

Fig. 3.10 (a) Cross-correlation of the control pulse out with the reference pulse for an inverter with a switching energy of 1 pJ and a fan-out of 28. (b) Autocorrelation of the control pulse emerging from the inverter.

frequency shifts down. We plot in Fig. 3.11 the measured spectral shift $\delta\nu$ normalized to the original spectral width $\Delta\nu_0$ versus the control power P_c/P_1 for propagation through only the second fiber. The spectral shift has an almost diode-like characteristic, with the knee around $P_c/P_1 \sim 0.5$. This behavior comes from the strong dependence of SSFS on pulse width ($\delta\nu \propto \tau^{-4}$), and the output pulse width versus control power is also plotted in Fig. 3.11.

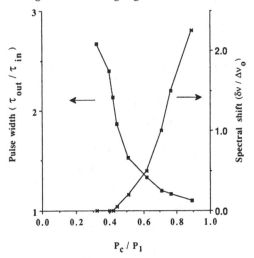

Fig. 3.11 Spectral shift $\delta\nu$ normalized to the original spectral width $\Delta\nu_0$ (right axis) and broadening of the control pulse $\tau_{\text{out}}/\tau_{\text{in}}$ (left axis) in the 2 km long polarization-maintaining fiber as a function of the input power P_c/P_1.

Although broader pulses reduce the soliton self-frequency shift, we still prefer to have $\tau_{\text{out}}/\tau_{\text{in}} \sim 1$ because of cascadability and for improved contrast in the cross-correlation. Despite the adiabatic broadening, the control pulse can be recompressed at the output by introducing a third fiber. From perturbation theory (cf. Appendix A), we know that the asymptotic value of the soliton is $(1+2a)\operatorname{sech}[(1+2a)t]$ when we start with an input electric field $(1+a)\operatorname{sech}t$ for $-0.5 < a < 0.5$. We can recompress an $a < 0$ pulse either by amplifying the output and then coupling into another fiber or by going straight into another fiber that has a lower soliton energy. If the output from the long fiber enters another fiber with normalized amplitude $(1+G)(1+2a)$, where $(1+2G) = (1+2a)^{-1}$, then the pulse shrinks back to its original pulse width in a few soliton periods. For example, we have done computer simulations for $-0.25 < a < -0.2$ and find that $\tau_{\text{out}}/\tau_{\text{in}}$ returns to 1 within 2 to 2.5 soliton periods.

Extension of time-domain chirp switch to semiconductors

As further confirmation of the mechanisms in the TDCS, we can replace the moderately birefringent fiber with an AlGaAs semiconductor waveguide as the nonlinear chirper. We use the semiconductor below half the energy gap to avoid two-photon absorption,

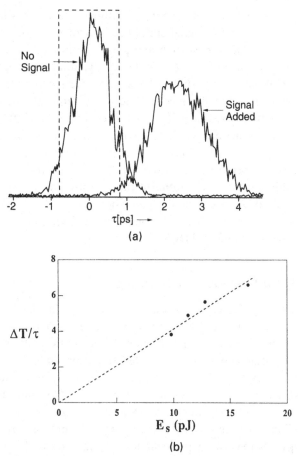

Fig. 3.12 (a) Time-shift keyed data for the hybrid time-domain chirp switch with a switching energy of 9.8 pJ and a fan-out of three. (b) Shift of the control pulse versus the signal energy for a control energy of 96.5 pJ in the waveguide.

and the instantaneous nonlinear index is more than two orders of magnitude larger than in fibers. Details of the waveguide structure are given in Refs [3.13, 3.14, 3.15] and further discussed in Section 4.2. The hybrid TDCS, which consists of a 2.1 mm long AlGaAs waveguide followed by a 600 m length of polarization-maintaining fiber (see Fig. 4.1), is tested in the same experimental apparatus as described above. A more detailed treatment of the physics and mechanisms in the hybrid TDCS is included in Section 4.2.

The time-shift keyed data for the hybrid TDCS is illustrated in Fig. 3.12 (a), where the signal energy in the waveguide is 9.8 pJ and the control energy is 96.5 pJ. The rectangle outlines the clock window, and we see that adding the signal shifts the control pulse out of this window. Because of mode mismatch and poor coupling into the fiber, the control energy exiting the fiber is 30.2 pJ, yielding a device fan-out or gain of about 3. We should be able to reduce the signal or switching energy closer to the picojoule level by reducing the cross-sectional area of the waveguide. The measured temporal shift of the control pulse as a function of signal energy is also included in Fig. 3.12 (b), and we expect the shift to be linearly proportional to the switching energy. Therefore, we have shown that by understanding the TDCS architecture we can optimize the performance of each section and extend to materials other than fiber. The hybrid TDCS is just a first step toward the desired goal of achieving a compact, integrable, all-semiconductor TDCS.

3.3 Billiard-ball soliton-interaction gates

Soliton-dragging logic gates utilize inelastic collisions between orthogonally polarized solitons to create a time shift. If instead temporal solitons are aligned along the same axis in optical fibers, then billiard-ball-like logic can be demonstrated based on elastic collisions between solitons. A temporal, conservative-logic interaction gate can perform AND, inversion and routing functions. Fredkin and Toffoli [3.16] introduced an interaction gate (Fig. 3.13 (a)) as a universal, conservative-logic primitive (i.e., a gate that does not, in principle, require energy dissipation to perform logic) and discussed a spatial implementation using the collision between billiard balls, where the logic operation corresponds to the locus of the balls (Fig. 3.13 (b)). The fundamental interest in the interaction gate results from the reversibility of the operation, although conservative gates may have limited practical value since generally they do not have logic-level restoration and they may not have gain or fan-out.

Solitons can be used to demonstrate billiard-ball models of logic because they are internally balanced pulses that behave in many ways like particles [3.17]. Because of the typically weak nonlinearity of most materials, the spatial interaction gate of Fig. 3.13 may be mapped into a temporal analog (i.e., various positions correspond to various time slots), which enables us to use low energy pulses in long lengths of optical fibers. Consider two initially separated solitons that are traveling toward each other in a fiber and have different group velocities or

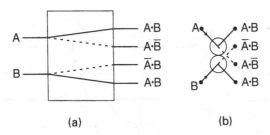

Fig. 3.13 (a) Schematic of an interaction gate and (b) a simple implementation using billiard balls [3.16].

center frequencies. Gordon [3.18] and Zahkarov and Shabat [3.19] have studied analytically the elastic collisions between these pulses and finds that: (a) although details in the interaction region depend on the phase between the two pulses, after the pulses separate the result of the collision is independent of the initial phase; (b) after the interaction the two pulses appear to pass through each other and return to their original velocities; and (c) as a result of the collision, both pulses are phase shifted and displaced, increasing their original separation.

Simulations for different phases between the pulses (Fig. 3.14) illustrate this unique collision property of solitons. A complete description of the collision equations and the two soliton function is included in Appendix C. The various cross-sections correspond to different points along the collision (i.e., varying position in the fiber), and Figs. 3.14(a)–(c) show the effects of varying the solitons' phase difference for equal amplitude pulses. Pulses that are in phase appear to attract, while pulses that are out of phase appear to repel each other. Although the behavior for $-10 \leqslant z/z_c \leqslant 10$ clearly depends on the phase between the pulses (the normalized length z_c is defined in Appendix A, and more detailed behavior in this range is given in Appendix C), after the pulses separate the output is independent of the original phase. In Fig. 3.14(d) the solitons have slightly different amplitudes, which shows that they can cross without ever seeming to merge. The dashed lines show the phase-independent asymptotic incoming and outgoing trajectories of the peak (or centroid) of the slower soliton. Since this is the asymptotic trajectory, the pulse centers align with the curve only outside the interaction region. We see that the interaction displaces each soliton away from the other. The parameters chosen for Fig. 3.14 closely match the experimental conditions discussed below; i.e., the solitons' displacement in the case of equal amplitudes is 3.4 times the pulse width.

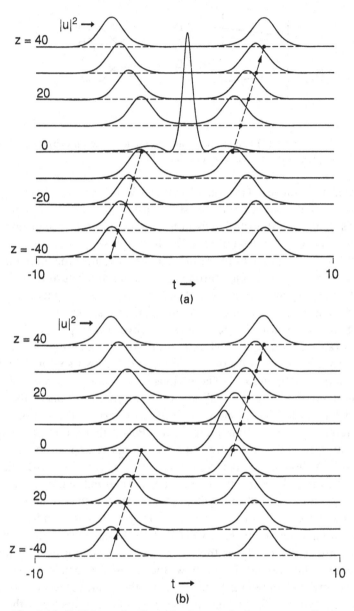

Fig. 3.14 Collisions of solitons having nearly the same amplitudes and frequencies, as discussed in Appendix C. The plots show $|u|^2$ as a function of t for values of z between -40 and $+40$, taken from the two-soliton function of Eq. (C.3). The frequencies of the two solitons are $\Omega = \pm0.05$ and the collision center is at $t = z = 0$. The frequency difference of 0.1 is about 9% of the half intensity width of the soliton's spectrum.

(*Continued on facing page.*)

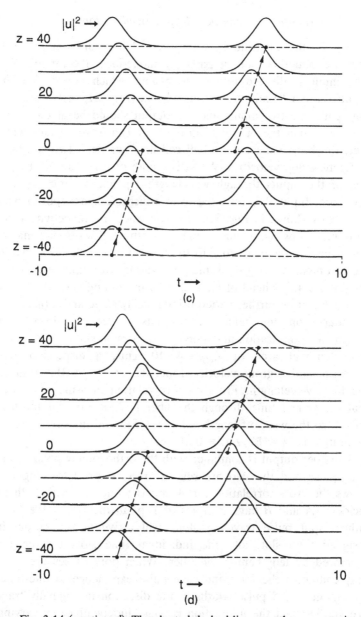

Fig. 3.14 (*continued*) The slanted dashed lines are the asymptotic incoming and outgoing trajectories of the slower soliton, showing its time displacement because of the collision. In plots (a), (b) and (c), the amplitudes A are both 1; the phase difference $\theta_1 - \theta_2$ at the center ($t = z = 0$) is 0, $\frac{1}{4}\pi$ and π, respectively. In plot (d) the amplitudes are 1.075 and 0.925, and the phase difference is π. After the collision the slower soliton remains the smaller one, so they appear to have crossed.

Experiments using parallel polarized solitons

In the experiments we use $\tau \sim 430\,$fs pulses at $\lambda \sim 1.67\,\mu$m that are obtained from a passively mode-locked color-center laser. The input is divided into three beams, which correspond to two inputs (A and B) nd one reference beam (Fig. 3.15(a)). For a cascadable gate, the two inputs and the output should be at the same frequency so that the output from one gate can drive an identical gate. Yet, to observe the billiard-ball collisions we need two solitons with different group velocities. We satisfy both of these criteria by passing one of the inputs through wavelength-dependent optics to skew its spectrum. Input A is reflected from a dichroic beam splitter with the reflectivity shown in Fig. 3.15(b), and since the reflectivity increases with increasing wavelength, the $\sim 10\,$nm (full-width-half-maximum) pulse spectrum appears to shift to longer wavelengths. The two inputs are recombined along the same polarization with input A (which travels slower) 4.5τ ahead of input B, and the timing and phase between A and B can be further varied using a delay stage and a piezo-electric transducer on one mirror. The inputs are coupled into a 1090 m length of polarization-maintaining, dispersion-shifted fiber with a cross-sectional area of $A_{\mathrm{eff}} \sim 4 \times 10^{-7}\,$cm^2, a dispersion zero at 1.584 μm and a group-velocity dispersion of $D \sim 4.81\,$ps/(nm km) at the laser wavelength. From cross-correlation measurements we find that the transit time through the fiber is 7.5τ different for the two pulses, so the walk-off length is $l_{\mathrm{wo}} \sim 145\,$m and the center frequency separation is $\delta\lambda = \delta T/DL \cong 0.61\,$nm.

The fiber output is combined with the orthogonally polarized reference pulse and the two are sent to a cross-correlator. Figure 3.16 shows the cross-correlation for A and B alone and with both pulses incident (A and B have energies of 17 pJ in the fiber). We find that either input pulse traveling alone will arrive at the output in its assigned time slot, and the individual time slots can be further separated by lengthening the fiber. When both pulses are incident, they collide in the fiber and each pulse experiences a one-time displacement of 3.5 pulse widths in the direction it originally traveled. We verified that the interaction is phase independent by varying the piezo-electric transducer and checking that the pulses remain well separated.

The logic operation performed depends on the time slot selected at the output. In time slot #1 we observe $A \cdot B$, which acts therefore as an AND-gate. In time slot #2 we observe $\bar{A} \cdot B$: if B is a clock pulse

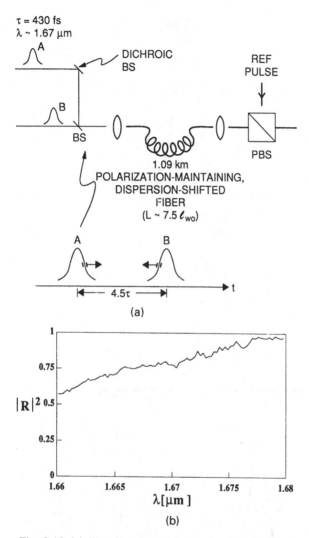

Fig. 3.15 (a) Experimental configuration for testing the soliton-interaction gate (BS = beam splitter, PBS = polarizing beam splitter). (b) Reflection coefficient of the dichroic beam splitter that is used to skew the spectrum to longer wavelengths of one input.

($B = 1$) then time slot #2 acts as an inverter. Furthermore, if A is a control beam, then time slots #1 and #2 appear as a temporal routing gate for B. We select the time slots centered around B since input B was not distorted by the dichroic beam splitter in this example. To cascade soliton-interaction gates, the desired time slot must be the

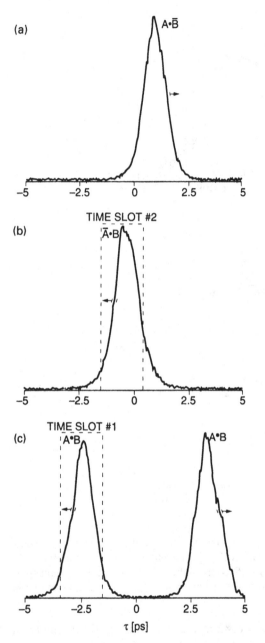

Fig. 3.16 Cross-correlation of soliton-interaction gate output with a reference pulse for: (a) A alone; (b) B alone; and (c) A and B incident. The two inputs A and B are 17 pJ each in the fiber.

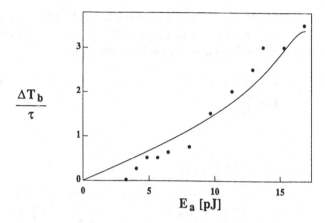

Fig. 3.17 Normalized displacement of input B versus energy of input A. The dots correspond to the experimental data, and the solid line is calculated from relations (3.4) and (3.5) with $E_b = E_s = 17\,\text{pJ}$ and $\Delta V = 0.1$.

only time slot seen by the next gate. For example, to cascade the AND-gate we align time slot #1 to the B-input of the next gate. Then, for a $7.5 l_{wo}$ fiber the next gate does not interact with the previous gate's $\overline{A} \cdot B$ or $A \cdot \overline{B}$ since they are at least 3.5τ farther away.

To test if the interaction gate can exhibit gain, we varied the amplitude of input A and measured the displacement of B with and without A. In Fig. 3.17 the dots represent the experimentally measured displacement, and we find that to obtain at least a 2τ shift requires about $11\,\text{pJ}$ of energy, which means that the gain is only 1.5. We study analytic formulas for the displacement of each pulse in order to understand how to improve the fan-out and reduce the switching energy. In Ref. 3.18 formulas for the displacement after the collision between two-color pulses are derived under the assumption that both pulses are solitons of the form $u_i = A_i \operatorname{sech} A_i t$. However, the inputs to our fiber are of the form $u_i = A_i \operatorname{sech} t$; i.e., we vary the amplitude for fixed laser-pulse width. Nonetheless, for a lowest-order approximation we can employ Gordon's formulas by replacing the amplitudes A_i by the normalized energies E_i/E_s, where E_s is the fundamental soliton energy and E_i ($i = a, b$) is the input energy. After all, it is the intensity times interaction length, which is proportional to the pulse energy, that determines the kick each pulse receives after colliding.

In the limit that the two pulses start separate and walk completely through each other, we find from relation (7) of Ref. 3.18 that the temporal shift of pulse B as a result of the collision is

$$\frac{\Delta T_b}{\tau} \cong \frac{1}{1.76\frac{E_b}{E_s}} \ln(X^2 + Y^2), \tag{3.4}$$

where

$$X = \frac{\left(\frac{E_b}{E_s}\right)^2 - \left(\frac{E_a}{E_s}\right)^2 + (\Delta V)^2}{\left(\frac{E_b}{E_s} - \frac{E_a}{E_s}\right)^2 + (\Delta V)^2}, \quad Y = -2\frac{\frac{E_a}{E_s}\Delta V}{\left(\frac{E_b}{E_s} - \frac{E_a}{E_s}\right)^2 + (\Delta V)^2}. \tag{3.5}$$

We note that the displacement can be increased by decreasing the normalized velocity difference ΔV or increasing the walk-off length. For these experiments the soliton period is given by $Z_0 = 0.322\pi^2 c\tau^2/\lambda^2|D| = 13.14\,\text{m}$ and the fundamental soliton energy is $E_s \cong P_1\tau = \lambda\tau A_{\text{eff}}/4n_2Z_0 = 17\,\text{pJ}$, where, for fused silica fibers, we use $n_2 = 3.2\times10^{-16}\,\text{cm}^2/\text{W}$. In addition, the normalized velocity difference is $\Delta V = 2\times1.76Z_0D\,\delta\lambda/\pi\tau \cong 0.1$.

Using these experimental values for the soliton parameters in Eqs. (3.4–5), we generate the solid curve in Fig. 3.17 for $E_b = E_s = 17\,\text{pJ}$. When $E_a = E_b = E_s$ theory and experiment agree, since both pulses are launched as fundamental solitons. For $E_a < 5\,\text{pJ}$ ($A_a < 0.5$) the pulse is no longer a soliton, and it disperses even before the pulses meet. The experiment diverges from the theory since we cannot obtain complete walk-through with the dispersive wave. Between $5\,\text{pJ} < E_a < 15\,\text{pJ}$ the experiment oscillates around the theory probably because of the radiation field and non-soliton components of the pulse.

Just as in the soliton-dragging logic gates, we can also draw a spatial analog for the soliton-interaction gates (Fig. 3.18). Once again the trajectory of either particle alone is given by the dashed line. When both particles are incident, the particles appear to cross and the interaction pushes the pulses apart to increase their original separation. Then, the two particles follow trajectories parallel to their initial paths. All the translation occurs in the interaction region, and increasing the propagation afterwards does not amplify the switching as in SDLGs. So, unlike the SDLGs of Fig. 3.7, SIGs have higher switching energies and it is difficult to make arbitrarily large gains.

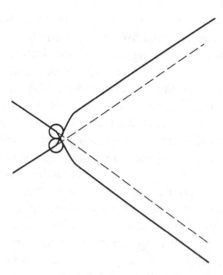

Fig. 3.18 Spatial analogy for soliton-interaction gates. The dashed lines are the trajectories for either pulse alone, and the solid lines are the trajectories with both pulses present.

Table 3.1 *Comparison of SDLG with SIG*

SDLG	SIG
• two temporally coincident, orthogonally polarized inputs	• temporally separate pulses along same axis
• pulse shifts frequency and ΔT from dispersive delay line	• one time displacement during interaction
• Switching energy ~ 1 pJ, fan-out up to 30 by lengthening fiber	• Switching energy $\gtrsim$ 11 pJ, fan-out < 1.5, increase by reducing $\delta\lambda$
• only use noninteracting pulse since unshifted frequency (inelastic collisions)	• use either shifted or unshifted output positions since same frequency (elastic collisions)
• polarizer at output blocks inputs and provides system cleanup	• collect "garbage" since same axis and overlapping spectra; need strobe

A more detailed comparison of SDLGs, which were discussed extensively in Sections 3.1 and 3.2, and SIGs is summarized in Table 3.1. In SIGs we use pulses that are along the same polarization axis and that are temporally separate at the input and output. Unlike the two-axis coupled nonlinear Schrödinger equations for a SDLG, the single axis equations describing soliton interaction gates are fully integrable, so the solitons undergo perfectly elastic collisions. Since the output pulses for a SIG have the same frequency content with or without a collision, either position of the pulse can be accepted at the cascadable output. However, the displacement accumulates only while the two inputs A and B are passing through each other, so the interaction length is limited to the walk-off length. Since all the inputs are along the same axis with overlapping spectra, soliton-interaction gates start to collect "garbage" in the unwanted time slots. Therefore, periodically we must use strobe gates such as SDLGs to select the values in the desired time slots and dump the remainder into an orthogonal polarization.

3.4 Soliton-trapping logic gates with output energy contrast

Within an ultrafast switching system we can use time-shift keying as the basis of logic. However, at the boundary of the system or the interface with electronics, we want to convert the time-shift-keyed signals to amplitude-shift keying. Soliton-trapping logic gates can perform the time-to-amplitude conversion at switching energies of $\sim 42\,\mathrm{pJ}$ and speeds comparable to the previously described gates. The physical mechanism for soliton trapping is cross-phase modulation between orthogonally polarized pulses just as in SDLGs; but soliton-trapping logic gates rely on frequency discrimination rather than temporal shifts. Soliton trapping is a special case of soliton dragging in which the two orthogonally polarized pulses are of comparable amplitudes and the fiber has a particular range of birefringences (see Appendix B).

Linear birefringence in optical fibers leads to pulse walk-off and broadening through polarization dispersion. However, Menyuk [3.20, 3.21] showed numerically that orthogonally polarized solitons can trap one another through cross-phase-modulation; thus, intensity-dependent effects can compensate for linear birefringence. For equal intensities along the two fiber principal axes, the two solitons shift their frequency in opposite directions, so that through group-velocity dispersion the soliton along the fast axis slows down, and the soliton along the slow axis speeds up. For a given birefringence and fiber length, a minimum intensity is required for the trapping.

To analyze the soliton-trapping mechanisms, we solved numerically the coupled nonlinear Schrödinger equation. As shown in Appendix B, if u and v are the electric fields along the two principal axes of the fiber, then the equations normalized in soliton units can be written as (t is local time on the pulse and z is normalized distance along the fiber):

$$-i\left(\frac{\partial u}{\partial z} + \delta\frac{\partial u}{\partial t}\right) = \frac{1}{2}\frac{\partial^2 u}{\partial t^2} + |u|^2 u + \tfrac{2}{3}|v|^2 u, \qquad (3.6\,(\text{a}))$$

$$-i\left(\frac{\partial v}{\partial z} - \delta\frac{\partial v}{\partial t}\right) = \frac{1}{2}\frac{\partial^2 v}{\partial t^2} + |v|^2 v + \tfrac{2}{3}|u|^2 v. \qquad (3.6\,(\text{b}))$$

The terms on the right-hand side correspond to group-velocity dispersion, self-phase modulation and cross-phase modulation. The pulses create an attractive potential well along the orthogonal axis through cross-phase modulation, and this coupling mechanism is independent of the phase between the two pulses. Birefringence is included through the term with $\delta = \pi\,\Delta n\,(\tau/1.76)/\lambda^2|D|$, where D is the dispersion in ps/(nm km) and Δn is the index difference between the two principal axes. A characteristic length for propagation in the fiber is the soliton period $Z_0 = 0.322\pi^2 c\tau^2/\lambda^2|D|$, and the fundamental soliton power is $P_1 = \lambda A_{\text{eff}}/4n_2 Z_0$, where A_{eff} is the effective core area. We solve the Eqs. (3.6) using a split-step Fourier transform technique [3.21] subject to the input conditions $u = A_x\,\text{sech}\,t$ and $v = A_y\,\text{sech}\,t$, where the soliton amplitudes are $A_x = \sqrt{P_x/P_1}$ and $A_y = \sqrt{P_y/P_1}$. It is convenient to define also the total amplitude $A_T = \sqrt{A_x^2 + A_y^2} = \sqrt{P_T/P_1}$ and the total power at the fiber input $P_T = P_x + P_y$.

In Fig. 3.19 we plot the intensity and spectrum obtained for parameters relevant to the experiments ($\delta \sim 0.517$, $L/Z_0 = 5.8$, anomalous dispersion $D > 0$) [3.9]. With a single pulse launched along the slow axis of the fiber ($A_x = 1.24/\sqrt{2}$, $A_y = 0$) the spectrum is symmetric, but the pulse moves with respect to $t = 0$ (Fig. 3.19 (a)) because of birefringence. In Fig. 3.19 (b) the input is split equally between the two axes ($A_x = A_y = 1.24/\sqrt{2}$, $A_T = 1.24$), and the two orthogonally polarized pulses trap near $t = 0$. The overlapping parts of the pulses (overlay of solid and dashed intensities in Fig. 3.19 (b)) narrow because of the increased intensity-dependent phase modulation, similar to soliton compression in fibers. The individual pulses

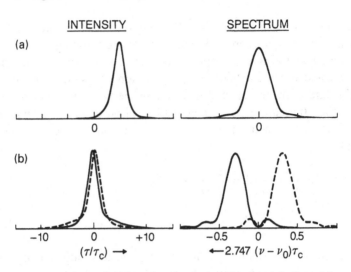

Fig. 3.19 Intensity and spectrum obtained by numerically solving Eqs. (3.6) for $\delta = 0.517$, $L = 5.8Z_0$ and (a) $A_x = 1.24/\sqrt{2}$, $A_y = 0$, or (b) $A_x = A_y = 1.24/\sqrt{2}$ and $A_T = 1.24$. The solid curve corresponds to the pulse polarized along the slow axis, and the dashed curve to the pulse polarized along the fast axis. The normalizing time is $\tau_c = \tau_p/1.76$, where τ_p is the full-width at half-maximum pulse width at the input.

are also asymmetric, with their wings extending farthest in the direction that polarization dispersion pulls the pulse. The spectra for the two pulses shift equally and in opposite directions so that the frequency difference compensates through group-velocity dispersion for the polarization dispersion. The pedestal or satellite bumps on the spectra correspond to the non-soliton part (for $A \neq 1$), which is stripped off through dispersion.

Spectral confirmation of soliton trapping

In the experimental apparatus of Fig. 3.20 $\tau \sim 300\,\text{fs}$ Gaussian pulses at $\lambda \sim 1.685\,\mu\text{m}$ are obtained from a passively mode-locked NaCl color-center laser. A variable attenuator is used to adjust the input power, and an isolator prevents feedback into the laser. The two polarizing beam splitters generate and recombine the two orthogonally polarized signals A and B, whose amplitudes are equalized using a half-wave plate; the two beams are separated, so we can easily block and unblock each arm. Another half-wave plate is used before the fiber to align the input polarization along the desired

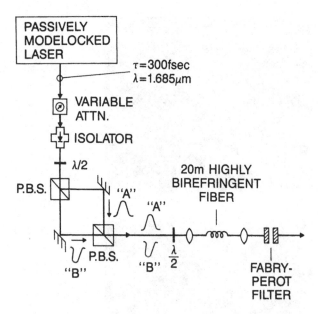

Fig. 3.20 Experimental apparatus for testing the soliton-trapping logic gates. *A* and *B* are the orthogonally polarized inputs. (PBS = polarizing beam splitter, $\lambda/2$ = half-wave plate).

fiber axes. The 20 m length of fiber used has a polarization dispersion of 80 ps/km ($\Delta n \sim 2.4 \times 10^{-5}$), a zero-dispersion wavelength of 1.51 μm, and a dispersion slope of 0.05 ps/(nm^2 km). By carefully winding the fiber on a 30 cm drum to minimize mode-coupling effects, we obtained polarization extinction ratios of $\sim$ 20:1.

Figure 3.21 illustrates the spectral confirmation of soliton trapping. When a pulse of $\sim$ 42 pJ energy propagates along a principal axis of the fiber, we obtain the output spectrum in Fig. 3.21 (a). The bifurcated spectrum of Fig. 3.21 (b) results with two identical input pulses of $\sim$ 42 pJ along each axis. For this same case, if we align a polarizer at the fiber output along the slow axis, we find the spectrum corresponding to a single trapped pulse (Fig. 3.21 (c)). Comparison between Figs. 3.21 (a) and 3.21 (c) shows that the solitons trap each other by shifting in wavelength and asymmetrizing. The spectrum in Fig. 3.21 (c) resembles closely the predicted spectrum from Fig. 3.19. For the fiber dispersion of $D \sim 8.75$ ps/(nm km) and the maximum separation of the spectral peaks in Fig. 3.21 (b) of $\sim$ 1 THz, we find $D \Delta\lambda \sim 83$ ps/km, which is close to the polarization dispersion of 80 ps/km measured for this fiber.

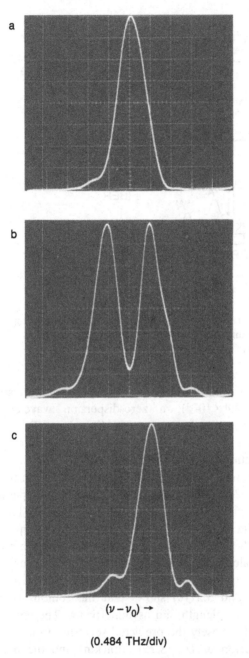

$(\nu - \nu_0) \rightarrow$

(0.484 THz/div)

Fig. 3.21 Spectral confirmation of soliton trapping in a fiber with a
polarization dispersion of 80 ps/km. (a) Pulse of ~ 42 pJ along a principal axis; (b) pulse of ~ 84 pJ at $\theta = 45°$, corresponding to two
equal amplitude pulses along both axes; and (c) same as (b), with a
polarizer at the fiber output aligned with the slow axis.

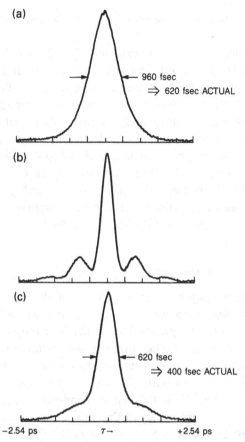

Fig. 3.22 Autocorrelations corresponding to the spectra of Fig. 3.21.
(a) Pulse of ~ 42 pJ along a principal axis; (b) pulse of ~ 84 pJ at
$\theta = 45°$; and (c) same as (b), with a polarizer at the fiber output
aligned with the slow axis.

The autocorrelation traces corresponding to the spectra of Fig.
3.21 are shown in Fig. 3.22. The ~ 42 pJ pulse traveling along a principal axis of the fiber broadens from 300 fs to 620 fs (Fig. 3.22 (a))
because the power is lower than that of an $N = 1$ fundamental soliton. With equal amplitude pulses along the two fiber axes (~ 84 pJ
net energy), the autocorrelation in Fig. 3.22 (b) shows interference
corresponding to the frequency separation of the two spectral peaks.
The trace in Fig. 3.22 (c) results by using a polarizer to look at the
trapped soliton along one axis of the fiber. As expected from Fig.
3.19, the center of the pulse narrows down to 400 fs, and the pedestal
corresponds to the tails on the pulse. Recall that an autocorrelation is
by definition symmetric, although the asymmetric spectrum in Fig.

3.21 (c) suggests an asymmetric temporal profile for the pulse of Fig. 3.22 (c).

For a polarization dispersion of 80 ps/km and $L = 20\,\text{m} = 5.8Z_0$, we find the maximum frequency splitting at a peak power of 280 W (energy $\sim 84\,\text{pJ}$). To obtain the fundamental soliton power P_1, we measured the pulse width as a function of power for propagation along a principal axis in a $L = 76\,\text{m} = 20Z_0$ length of fiber, and compared the results with soliton propagation calculations. From these measurements we extract a peak fundamental soliton power of $P_1 = 182\,\text{W}$ (energy $\sim 54.6\,\text{pJ}$). Therefore, the trapping for $\delta \sim 0.517$ occurs at $A_T \sim 1.24$, which are the values used to calculate Fig. 3.19. This agrees with Menyuk's calculations [3.21], where trapping occurs for $A_T = 1.2$ when $\delta = 0.5$ and $L = 5Z_0$ (see Appendix B).

Logic gates with frequency filter

By adding a frequency filter at the output of the fiber, the soliton-trapping device becomes a logic gate [3.22]. We now add an adjustable Fabry–Perot as the frequency filter at the fiber output with sufficient free spectral range so another order is not included within the pulse bandwidth. To obtain a high contrast ratio in an inverter or exclusive-OR gate, a Fabry–Perot with 85% reflecting mirrors (maximum finesse of 20) was used with the central-band-pass frequency adjusted to the original center frequency of the pulses. Figure 3.23 (b) shows an exclusive-OR gate with $\sim 8\!:\!1$ contrast ratio. We obtain a large output with either A or B alone and a factor of 8 reduction in spectral intensity when both A and B are present. By placing a polarizer at the fiber output along the principal axis corresponding to B, the exclusive-OR converts to an inverter: if B is left on, then we obtain NOT A. Although the input and output center frequencies of the device coincide, the output pulses broaden to approximately the inverse of the frequency band pass $\tau_{\text{out}} \sim 1/\Delta\nu \sim 4\,\text{ps}$. It is difficult to restore these output pulses to their original width and shape.

The inverter can become a cascadable gate if we widen the filter band pass so that the output pulses can propagate as solitons in a fiber. With 70% reflecting mirrors in the Fabry–Perot filter, a band pass of $\sim 0.58\,\text{THz}$ was obtained, and the central band pass was again set at the original pulse-center frequency. The lower finesse of the filter reduces the contrast ratio of the logic gate. Figure 3.24 shows the spectra and autocorrelations of the output from an inverter (we use a polarizer at the fiber output along B polarization, and treat A

"B" ALONE   "A" & "B" INCIDENT

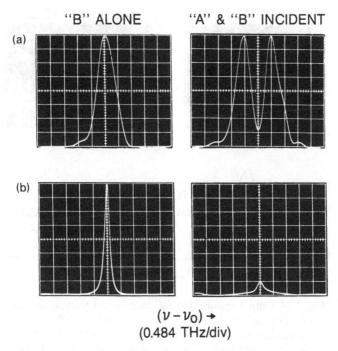

$(\nu - \nu_0) \rightarrow$
(0.484 THz/div)

Fig. 3.23 (a) Spectra direct from the fiber output. (b) Spectra after a Fabry–Perot band pass filter with 85% reflecting mirrors. This corresponds to an exclusive-OR gate with an ~ 8:1 contrast ratio.

as the signal). The contrast ratio is ~ 5:1 in the autocorrelation, ~ 4:1 in the peak spectral intensity, and ~ 3.2:1 in the net energy output. With no A signal, the 620 fs pulses out of the fiber broaden to 930 fs after passage through the filter. When both A and B are incident on the fiber, the output has a lower intensity, and the trapped solitons are narrower in the center. Although autocorrelations do not reveal absolute timing of the pulses, the B alone signal comes out at a different time from the trapped solitons.

By placing a band-pass filter at the original pulse-center frequency, the soliton trapping gates show inversion and exclusive-OR functionalities. However, if the filter pass band is centered ~ 0.5–1 THz away from the original pulse frequency, then we have an AND-gate: i.e., the frequency shifts out only when both pulses are present [3.23]. Figure 3.25 shows the spectra directly at the output of the fiber along the slow axis for a single pulse (A alone) and for a pair of temporally coincident pulses along each axis (A and B incident). Each of the pulses is ~ 42 pJ in energy, which is about three-quarters of the $N = 1$ fundamental soliton energy. This curve is similar to Fig.

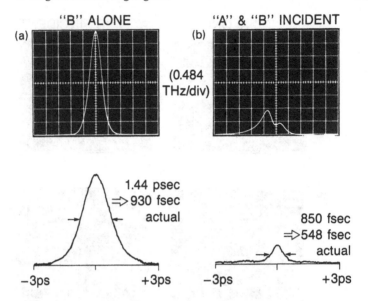

Fig. 3.24 Output spectra and autocorrelations from an inverter
(NOT A) with 70% reflecting mirrors in the Fabry–Perot filter, and
a polarizer along the B axis. (a) A blocked and (b) A unblocked.
Both spectra and both autocorrelations are plotted on the same
scale.

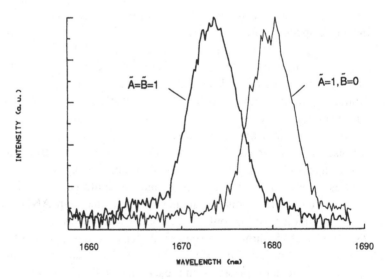

Fig. 3.25 Digitized spectra at the fiber output after a polarizer for a
single and a pair of pulses propagating in the fiber.

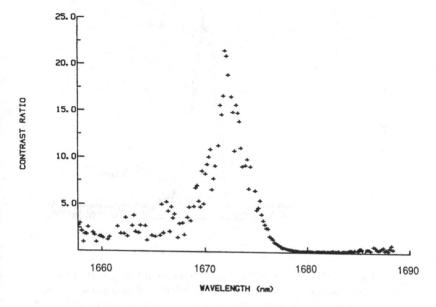

Fig. 3.26 Expected contrast ratio for the soliton-trapping AND-gate obtained by dividing the two curves (*A* AND *B* over *A* alone) in Fig. 3.25.

3.21 except that now a polarizer is placed at the fiber output to select only the slow axis signal, and the data is sampled and digitized for analysis. We determine the optimum center wavelength for the pass band of the frequency filter by dividing the signal for *A* and *B* by the signal for *A* alone. Figure 3.26 represents the expected contrast ratio as a function of wavelength, indicating that the contrast should exceed 20:1 for a center wavelength of the frequency filter of ~ 1672 nm. The contrast ratio between 1660 nm and 1670 nm appears noisy because of division by the small value of the *A* and *B* signal.

The soliton-trapping AND-gate (STAG) is implemented with 85% reflecting mirrors (maximum finesse of 20) in the Fabry–Perot filter, and Fig. 3.27 shows the spectra at the output of the STAG. The maximum contrast is ~ 22:1 at 1672 nm, which agrees with predictions, and the energy output for the on-state is ~ 9.9 pJ. The polarizer introduces a 3 dB loss, which should not be fundamental, and the transmission through the Fabry–Perot adds another 3 dB of loss. Since the frequency filter with a pass band narrower than the pulse spectral width attenuates and broadens the pulse, this device is not cascadable because both the center wavelength and the pulse shape at the output are different from the input.

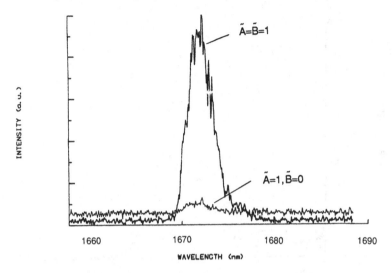

Fig. 3.27 Experimentally measured spectra at the output of the AND-gate, which consists of the fiber followed by a polarizer and Fabry–Perot frequency filter. The peak contrast ratio is 22:1, which agrees well with the values in Fig. 3.25. Note that we displace the origin of the two spectra simply for clarity.

3.5 Cascade of soliton-dragging and -trapping gates

In Section 5.3 I will describe the design of a 100 Gbit/s, self-routing packet, soliton ring network that uses ultrafast gates to select and decode the header [3.24, 3.25]. The code-matching logic module, which consists of four soliton logic gates, operates at the bit-rate and provides electronic outputs to control a network of routing switches. In the simplest protocol, the code-matching logic decides if the packet is EMPTY or if it has arrived at its destination and should be READ. We reserve an 8-bit header with all ones to designate an empty packet, so generating the EMPTY signal requires only a NOR-gate followed by an AND-gate. We implement the EMPTY signal circuit by cascading an SDLG with an STAG, and we measure an energy contrast of ~ 12:1 from the pair. Therefore, we use the cascadability and fan-out of SDLGs for operations that require several levels of logic and use the energy contrast from STAGs as the interface to electronics.

Figure 3.28 shows the experimental configuration for testing the cascade of logic gates, and the inset provides a schematic of the logic circuit. A passively mode-locked color-center laser ($\tau \sim 330\,\mathrm{fs}$ and $\lambda \sim 1.695\,\mu\mathrm{m}$) provides the control, signal and clock pulses. The

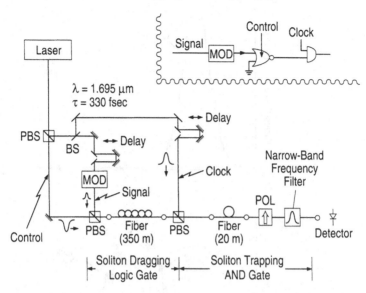

Fig. 3.28 Experimental apparatus for testing the cascaded soliton-dragging and -trapping gates (MOD = modulator, POL = polarizer, BS = beam splitter, PBS = polarizing beam splitter). The insert shows a schematic of the logic circuit.

timing for the various arms is adjusted so that the control and signal are temporally coincident at the input to the SDLG, while the control and clock overlap at the input of the STAG in the absence of the signal. The SDLG consists of a 350 m length of moderately birefringent fiber (polarization dispersion of ~ 90 ps/km) surrounded by two polarizing beam splitters (cf. Section 3.1). The STAG is implemented in a 20 m length of moderately birefringent fiber followed by a polarizer and a narrow-band frequency filter (cf. Section 3.4). By carefully handling the fibers to minimize mode-coupling effects, we maintain polarization extinction ratios in excess of 15:1 throughout. The signal is modulated with a chopper, and the output of the pair of gates is detected with a germanium detector.

In the SDLG, the control pulse is aligned with the fast axis and the signal is aligned with the slow axis. Because of soliton self-frequency shift (SSFS) effects, the control pulse shifts slightly to longer wavelengths after traveling through the SDLG fiber; for anomalous group-velocity dispersion, longer wavelengths travel slower. Consequently, we align the control pulse with the fast axis in the STAG fiber so that the SSFS partially cancels the velocity difference due to birefringence and increases the interaction length. The clock pulse, which is along the slow axis, speeds up due to soliton trapping by shifting to shorter

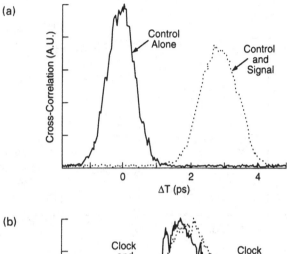

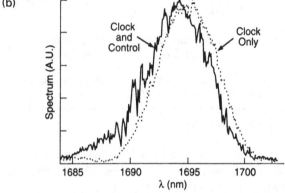

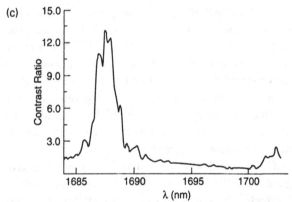

Fig. 3.29 (a) Cross-correlation of clock and control pulses at the output of the soliton-dragging logic gate with and without the signal pulse (E_{CONTROL} = 49 pJ, E_{CLOCK} = 44 pJ, E_{SIGNAL} = 10 pJ). (b) Spectrum at the output of the soliton-trapping AND-gate along the clock axis with and without the control present (E_{CONTROL} = 25 pJ, E_{CLOCK} = 44 pJ in the second fiber). (c) Ratio of the two spectra in (b) that shows a peak contrast of ~ 13:1.

wavelengths. Therefore, by placing the output polarizer along the clock axis and monitoring the up-shifted frequency, we avoid erroneous readings from SSFS or cross-Raman effects (cf. Appendix B).

As explained in Sections 3.1 and 3.2, in the SDLG the propagation time for the control changes when the signal is added. For example, Fig. 3.29 (a) shows the cross-correlation of a 49 pJ control pulse and a clock pulse at the output of the SDLG device, and we can shift the control by ~ 3 ps by adding a 10 pJ signal pulse at the input. For the 330 fs pulses we measure an $N = 1$ soliton energy of $E_1 \sim 73$ pJ, and, therefore, we find that the control pulse broadens after the first fiber by a factor of 1.6. In soliton trapping, two temporally coincident, orthogonally polarized solitons shift their frequencies to compensate for the polarization dispersion, as described in detail in Section 3.4. Figure 3.29 (b) shows the clock spectrum after the polarizer for a 44 pJ clock pulse with and without the temporally coincident control pulse, which is only 25 pJ in the STAG fiber due to coupling losses between the two gates. The spectral shift in Fig. 3.29 (b) is not as pronounced as in Fig. 3.25 because the control energy is smaller than used in Section 3.4 and because the control pulse input to the STAG is temporally broader by 60% than the clock input. In Fig. 3.29 (c) the ratio of the spectra shows a peak contrast of $\sim 13{:}1$.

By adding a narrow-band frequency filter after the soliton-trapping fiber, we obtain an AND-gate that has a high output only when both STAG inputs are temporally coincident. Figure 3.30 shows the detector output with the filter bandpass at 1688 nm in the tail of the clock spectrum and the signal modulated (the zero level is at the bottom of the scale). Thus, the measured contrast ratio is $\sim 12{:}1$ after the cascade of the two gates. When the signal is ON, the control shifts out of the clock window, and the clock propagates unperturbed in the STAG fiber. On the other hand, if the signal is blocked, then the clock and control coincide at the STAG input, and the clock spectrum shifts toward the filter bandpass. The "1" state is noisier because of jitter between the gates, while the "0" state corresponds to the clock alone. Although the contrast ratio from a STAG alone can be as high as 22:1 (Section 3.4), the result is degraded because of timing jitter and coupling losses between the two gates. Since the soliton ring network is designed for an 8-bit header, a contrast of at least 8:1 is required to avoid errors.

Thus, the cascade of gates in Fig. 3.28 represents the simplest configuration of multi-level ultrafast digital soliton logic gates that is needed for the header reader in a node of a 100 Gbit/s soliton ring

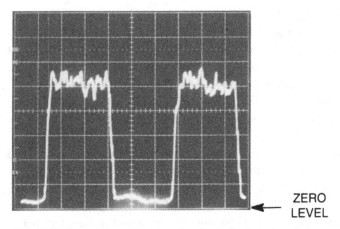

ZERO
LEVEL

Fig. 3.30 Output from the cascade of soliton-dragging and -trapping gates after a frequency filter centered at 1688 nm. The signal is modulated at a 50% duty cycle, and the contrast is ~ 12:1 (zero at bottom of scale).

network. The soliton-dragging NOR-gate has fan-out and a time-shift keyed output, while the soliton-trapping AND-gate converts the time shift to an energy contrast. Even with timing jitter and coupling losses the measured contrast at the output is ~ 12:1.

3.6 Summary

In summary, three types of ultrafast digital logic gates have been shown based on cross-phase modulation and soliton interactions in optical fibers. Logic gates are regenerative switches in which logic-level restoration is achieved by replacing the input photons with new photons from the power supply. We use solitons in fibers to avoid deleterious effects from group-velocity dispersion and nonlinearity and because fundamental solitons have particle-like properties.

Soliton-dragging logic gates satisfy all requirements for a digital optical processor such as fan-out, cascadability and Boolean completeness. In soliton dragging, the speed of travel through a fiber depends on whether a single pulse or a pair of pulses are present. Two orthogonally polarized, temporally coincident solitons undergo an inelastic collision, and as a result of cross-phase modulation and walk-off, the center frequencies of the pulses are altered. SDLGs are one example of a TDCS architecture in which a soliton-dispersive delay line acts as a "lever arm" to magnify small changes in center frequency. By separating the nonlinear chirper from the time and phase shifting needed for switching, the TDCS architecture reduces the requirements on the nonlinear material and leads to low switching energy for short pulses.

Soliton-interaction gates use elastic collisions between solitons along the same principle axis of a fiber. Solitons are remarkable in that although the details during the interaction depend on the phase between pulses, after the pulses separate their behavior is independent of the initial phase. SIG's demonstrate that the particle-like properties of solitons can implement a conservative, billiard-ball logic. The spatial analogies of Figs. 3.7 and 3.18 help to compare the SDLGs and SIGs: in SDLGs most of the shift accumulates while propagating at a new angle, as contrasted with SIGs, where the shift occurs only during the interaction. Consequently, in SIGs it is difficult to obtain large gains or lower switching energies.

Soliton trapping is a special case of soliton dragging when two pulses are of comparable amplitude and there is a particular range of birefringence. In addition, soliton trapping is an example of how solitons can counteract the effect of birefringence because of their internal restoring force. We demonstrated a soliton-trapping AND-gate that is sensitive to the timing of the input pulses and has an output with a contrast ratio $> 20:1$. Although STAG's do not have gain or cascadability, they can be used as the final stage in an all-optical switching system to convert the time-shift keying to an amplitude-shift-keyed output compatible with electronics. As an example, we showed a cascade of an SDLG and an STAG whose output is measured with a detector to have a contrast of $\sim 12:1$.

In Chapter 1 the key challenges in implementing all-optical devices were reviewed, and the devices in this chapter illustrate approaches to tackle each of those problems. We make a three-terminal device by using the two orthogonal polarized inputs and a spatially separated output. To guarantee cascadability, both inputs and the output are at the same optical frequency and we use solitons, which can retain their pulse shape and width while propagating through the fiber. Phase-independent operation is achieved by using cross-phase modulation in fibers, which depends only on the intensities and not on the phase of the electric fields of the orthogonally polarized signals. To obtain fan-out without using an amplifier, the temporal axis is used in the switching operation. For example, the logic is based on time-shift keying, and logic operations correspond to moving a pulse in and out of a clock window. SDLGs operate at switching energies approaching a picojoule by using a dispersive delay line as a "lever arm," and the concept is elucidated by the time-domain chirp switch architecture. Techniques for reducing timing jitter and relaxing the timing constraints of terabit systems are described in the next chapter. Also, synchronization of power supply lasers by using an optical phase-locked loop is described in Ref. 3.25.

4

Timing constraints in terabit systems

As the bit rates increase and the spacing between bits decreases, the timing jitter tolerances in high-performance systems become increasingly tight. For example, in a 0.2 Tbit/s system that uses 500 fs pulses with a 10% duty cycle, the timing at each gate must be within several pulse widths. Note that 500 fs corresponds to 150 μm in free space or about 100 μm of optical fiber. Optical fibers have an inherent advantage over copper wires in that the velocity of a signal does not depend on the adjacent signals. Even if we carefully adjust the delays throughout the optical system at a given set of conditions, there will be timing jitter introduced from environmental effects such as thermal fluctuations, laser phase noise, and nonlinear effects that couple intensity to timing variations. There are several approaches to relaxing the timing constraints of terabit-rate systems, and, to be specific, we will concentrate on a digital optical system based on soliton-dragging logic gates (SDLG), which were described in Section 3.1.

From the device viewpoint, there are at least two different approaches for combating the timing jitter problem in the synchronous system. First, the timing of the input bit streams can be corrected with respect to a standard clock signal. Second, we can relax the timing restrictions of the logic gates by increasing the input clock window accepted by the gates. For example, we saw this second approach applied to a nonlinear optical loop mirror in Section 2.4 (also see Section 5.1), where orthogonally polarized pulses slip completely through each other.

Clock skew and synchronization are also major concerns for terabit-rate systems. The clock pulses must be properly distributed so that all gates operate in phase with the clock cycle, and all the lasers in the system must be synchronized. Furthermore, each input to the system must enter at the correct point in the clock cycle. One approach for solving some of these synchronization issues is discussed in Ref. [4.1] and Section 5.3. In particular, all lasers are operated at a uniform repetition rate to avoid the bit-level synchronization required when, for example, combining bit streams from different locations in the network. Therefore, the problem of triggering within a bit period reduces to matching sinusoids in periodicity and phase. Optical phase-locked loops [4.2] can solve this problem, since the bandwidth of the feedback loop needs to be the bandwidth of the noise rather than the bandwidth of the data stream.

In Section 4.1 general formulas for length and energy restrictions pertaining to SDLGs identify various sources of timing jitter. All-optical timing restoration using a hybrid time-domain chirp switch (cf. Section 3.2) is described in Section 4.2. Then, two techniques for relaxing the timing of SDLGs are discussed: (a) in Section 4.3 using a passive prechirper; and (b) in Section 4.4 using an active erbium-doped fiber amplifier (EDFA). Although the examples are specific to SDLGs, the concepts should be applicable to any high-speed switching or transmission system.

4.1 Energy and length formulas for soliton-dragging gates

To understand limitations on SDLGs and sources of timing jitter, I review some fundamental relations for soliton dragging. More generally, these limitations relate to picosecond and femtosecond pulse propagation in optical fiber systems. As described in Section 3.2, SDLGs are one example of a more general time-domain chirp switch (TDCS) architecture, which consists of a nonlinear chirper followed by a soliton-dispersive delay line [4.3]. In particular, for an SDLG the nonlinear chirper corresponds to a moderately birefringent fiber (fiber #1) and the dispersive delay line corresponds to a length (typically $\sim 300\,\text{m}$) of polarization-maintaining fiber (fiber #2).

Because of power supply limitations, it is important to minimize the switching energy. For a cascadable gate, the control or power supply pulse should correspond to an $N = 1$ fundamental soliton that has an energy (Appendix A)

$$E_{\text{c}} = \frac{1.76}{2\pi^2} \frac{\lambda_0^3 A_{\text{eff}}^{(2)} D^{(2)}}{n_2^{(2)} c \tau} , \tag{4.1}$$

where τ is the pulse width, A_{eff} is the effective cross-sectional area, D is the group-velocity dispersion and the superscripts (1)/(2) correspond to the moderately birefringent/polarization-maintaining fibers. In addition, n_2 is the nonlinear index of refraction, λ_0 is the wavelength in free space and c is the speed of light. For the two-fiber TDCS, the signal or switching energy is given by [3.4, 4.5]

$$E_s = \frac{\Delta T}{\tau}\, \tau^2\, \frac{\Delta n^{(1)}}{L^{(2)} D^{(2)}}\, \frac{A_{\text{eff}}^{(1)}}{n_2^{(1)}}\, \frac{3}{1.76\lambda_0 B}\,, \qquad (4.2)$$

where Δn is the birefringence and the cross-phase modulation coefficient B is $\frac{2}{3}$ for a linearly birefringent fiber. Based on the details of the system, $\Delta T/\tau$ will in general be a constant (i.e., we require a certain amount of shift to move the control pulse out of the clock window). There are two reasons why E_s is proportional to τ^2: (1) the shorter the pulse width, the less the pulse has to be moved; and (2) we require a particular intensity for the nonlinear effect, and the energy is τ times that intensity. Furthermore, if we assume that a 10τ separation is required to keep adjacent solitons from interacting, then the maximum bit rate is

$$\text{bit rate}|_{\text{max}} \sim \frac{1}{10\tau}\,. \qquad (4.3)$$

Therefore, for the minimum switching energy and maximum bit rate, we want to use as short a pulse as possible, although the penalty is an increase in the control or power supply energy. However, the minimum τ is limited by soliton self-frequency shift (SSFS) [4.6], and we find experimentally that if $L^{(2)} \sim 300\text{--}500\,\text{m}$ then the shortest usable pulse width is $\tau \sim 0.5\,\text{ps}$. We also want to minimize the birefringence in the first fiber, although we need to maintain polarization extinction to avoid linear interference between the control and signal. Finally, the simplest way of reducing E_s is to lengthen $L^{(2)}$ as far as possible (cf. Section 3.2).

The length $L^{(2)}$ that can be used in an SDLG is constrained at least by four factors: loss (L_α), latency (L_{lat}), SSFS or other higher-order nonlinear effects (L_{ssfs}) and thermal stability (L_{th}). The maximum allowable length in a real system application will be the shortest of the above four lengths. First, the loss length is just the 3 dB loss length, or $L_\alpha \sim 1/\alpha$. For standard fused-silica fibers with a loss of 0.14 dB/km, we find that $L_\alpha \sim 21\,\text{km}$, which is generally much longer than the other lengths. Second, based on the application and the system architecture, there may be a maximum pipeline depth that is allowed, where the pipeline depth is the latency divided by the bit period (i.e., the number of bits that fit in the device). For example, for current pipelined architectures the maximum depth is around 1000. However, for our typical gate lengths around 300 m and

0.2 Tbit/s data rate, the pipeline depth would be about 3×10^5. Since our interest is primarily in telecommunications, where the system can be designed to be feed-forward, we will assume that L_{lat} is not a limitation.

Because of SSFS (or, equivalently, a low frequency Raman gain) the control soliton shifts to longer wavelengths. The shift is a strong function of pulse width τ ($\Delta\nu \sim \tau^{-4}$) and scales linearly with intensity. SSFS must be avoided for two reasons: (1) the frequency shift ruins the cascadability of the gate; and (2) SSFS couples amplitude fluctuations to frequency and timing fluctuations. From Gordon's theory [4.6] (refer to Appendix A) we find for fused silica fibers that

$$\frac{d\nu}{dz}[\text{THz/km}] = \frac{0.0436}{\tau^4[\text{ps}]}\left(\frac{\lambda[\mu\text{m}]}{1.5}\right)^2 \frac{D^{(2)}[\text{ps/(nm km)}]}{15} \frac{P}{P_1},$$

(4.4)

where P_1 is the peak fundamental soliton power. From this equation we notice that intensity fluctuations lead to a frequency fluctuation that will also cause a timing jitter. To guarantee cascadability, we insist that the frequency shift $\delta\nu$ be less than one-sixth of the spectral width; i.e.,

$$\delta\nu < \tfrac{1}{6}\Delta\nu_0 \sim \frac{1}{19\tau},$$

(4.5)

where $\Delta\nu_0$ is the full-width half-maximum spectral width of the pulse (for hyperbolic secant pulses $\Delta\nu_0 \sim 0.315/\tau$). Rewriting relation (4.4) using relation (4.5), we find that the maximum length restricted by SSFS in fused silica is

$$L_{\text{ssfs}}[\text{km}] \sim 1.2\tau^3[\text{ps}]\left(\frac{1.5}{\lambda[\mu\text{m}]}\right)^2 \frac{15}{D^{(2)}[\text{ps/(nm km)}]} \frac{P_1}{P}$$

(4.6)

or $L_{\text{ssfs}} \propto \tau^3/\lambda^2 D^{(2)}$. As an example, for the SDLG experiments with $\tau = 0.5\,\text{ps}$, $\lambda = 1.685\,\mu\text{m}$ and $D^{(2)} = 5\,\text{ps/(nm km)}$ we calculate $L_{\text{ssfs}} \sim 358\,\text{m}$ when the control pulse is an $N = 1$ soliton, which is comparable to the 350 m fiber that was used in Section 3.1 [4.7]. Consequently, the demonstrated SDLGs are close to the length limit set by SSFS.

The thermal length L_{th} turns out to be even more restrictive in the SDLG experiments than the L_{ssfs}. A temperature variation affects the fiber both because of a change in length and because of a change in index [4.8]. However, the index variation is almost ten times stronger than the stretching effect in fibers, so we will ignore the latter. The temperature coefficient for fused silica glass fibers is $\partial n/\partial T = 1.14 \times 10^{-5}\,°C^{-1}$. If we assume that we can stabilize the temperature variations to within $\Delta T \sim \frac{1}{10}\,°C$, then the index fluctuation is $\delta n \sim 1.14 \times 10^{-6}$. The time shift associated with this index variation is $\hat{\delta t} = (l/c)\,\delta n$ and, if the acceptable time window is ξ pulse widths wide ($\hat{\delta t} = \xi\tau$), then the thermal length is $L_{th} = \xi c\tau/\delta n$. For the experiments that use $\tau \sim 0.5\,\text{ps}$ pulses we find that $L_{th} \approx \xi \times 132\,\text{m}$: thus, for a one pulse width time window we find $L_{th} = 132\,\text{m}$, for a 2τ time window we find $L_{th} = 264\,\text{m}$, etc. Although in a laboratory environment where data is taken over a short period of time we can use fiber lengths much longer than L_{th}, in real applications we must increase ξ to make L_{th} longer than the device length.

Even without considering other noise sources such as the laser phase noise, we find two major sources of timing jitter in the $\sim 300\,\text{m}$ SDLGs from SSFS and temperature. Thermal considerations dictate that the acceptable time window be more than 2τ. On the other hand, the maximum window can only be $\sim 5\tau$ since this is half of the 10τ separation between adjacent bits. The design value for the SDLG window should lie in this range and may be determined by other system constraints as well as whether the bit stream's timing is corrected, as discussed in the next section.

4.2 Timing restoration using a hybrid time-domain chirp switch

In optical switching and transmission systems it is important to periodically restore logic level and timing. Current networks perform the restoration at regenerators. However, with the trend toward "light-pipe" systems in which the regenerators are replaced by EDFAs, the system becomes limited by timing jitter and fluctuations. Here we show that a hybrid TDCS [4.9] that consists of a semiconductor waveguide followed by a polarization-maintaining fiber, as described in Section 3.2, can act as an ultrafast, all-optical timing restorer (i.e., we can reduce the timing jitter of the input bit stream).

The hybrid TDCS, which is schematically shown in Fig. 4.1, is tested in the same experimental apparatus as described in Sections 3.1 and 3.2. A passively mode-locked color-center laser supplies $\tau \sim 415\,\text{fs}$ pulses near $1.69\,\mu\text{m}$. The soliton delay line is $600\,\text{m}$ of polarization-maintaining, dispersion-shifted fiber with a zero-dispersion wavelength of $1.585\,\mu\text{m}$. The insert to Fig. 4.1 shows the

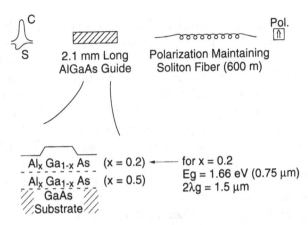

Fig. 4.1 Schematic of the hybrid time-domain chirp switch and details of the semiconductor waveguide.

details of the 2.1 mm long AlGaAs waveguide, which has a cross-sectional area of approximately 2.5 μm × 5 μm. The ridge waveguide is formed in a 2.55 μm thick layer of $Al_{0.2}Ga_{0.8}As$, and the guiding is assured by a 2.55 μm buffer layer of $Al_{0.5}Ga_{0.5}As$ that has a refractive index 0.15 less than the active layer. The material composition is chosen so that the laser spectrum lies more than 100 meV below the half-gap energy, thus avoiding two-photon absorption. In this wavelength range we find that $n_2 \sim 3.6 \times 10^{-14} cm^2/W$ and that the material is isotropic (e.g., cross-phase modulation is two-thirds of self-phase modulation) [4.10, 4.11, 4.12]. We obtain a π-phase shift from self-phase modulation with less than a 10% absorption and find that the nonlinear absorption originates primarily from three-photon absorption. Furthermore, time-resolved pump-probe measurements confirm that the nonlinearity is instantaneous on the 500 fs time scale of the pulses.

Unlike soliton-dragging where pulses walk apart because of the birefringence in the fiber, time-resolved measurements in semiconductor waveguides as long as 7.7 mm show negligible walk-off for the 415 fs pulses. Therefore, when the two orthogonally polarized pulses overlap in the semiconductor waveguide, they maintain their relative positions and experience the same intensity and slope throughout their interaction. To understand the implications of no walk-off, we studied experimentally the shift of the control pulse ΔT as a function of the initial separation between the control and signal pulses δt (Fig. 4.2). When the two pulses are perfectly overlapped there is no time shift: the cross-phase modulation just enhances the self-phase

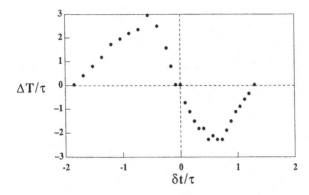

Fig. 4.2 Shift of the control pulse versus the initial separation between the control and signal pulses.

modulation, which leads to a symmetric broadening of the spectrum. On the other hand, when the pulses are partially separated and one pulse travels on the wing or side of the other pulse, there can be a net chirp or shift of the center frequency.

The data in Fig. 4.2 also shows that the hybrid TDCS can provide timing restoration. For example, suppose that the control pulse is the input pulse and that the signal pulse is a "reference" pulse with the proper temporal position. We define the initial separation between the pulses $\delta t = t_{\text{input}} - t_{\text{ref}}$ and the resulting shift of the input pulse ΔT. Therefore, if the input pulse is earlier than the reference pulse ($\delta t < 0$), then the nonlinear interaction pulls the input pulse to later times ($\Delta T > 0$), and vice versa.

We can intuitively understand the timing correction by considering the effects of frequency shift and anomalous group-velocity dispersion. The instantaneous frequency change $\delta \omega$ of the input pulse because of cross-phase modulation is given by [4.13]

$$\delta \omega(t) = -\frac{\partial \Delta \Phi}{\partial t} = -\frac{2\pi}{\lambda} \frac{2}{3} n_2 L \frac{\partial I_{\text{ref}}}{\partial t} \qquad (4.7)$$

and is proportional to the negative slope of the reference pulse for $n_2 > 0$. Since lower frequencies travel slower in anomalous dispersive material, a pulse with downshifted frequency arrives later and travels toward the right on a time axis. Figure 4.3 (a) illustrates the correcting nature of the interaction. When the input pulse arrives earlier than the reference pulse (top case), the input pulse sees a positive index slope, which lowers its instantaneous frequency, which in turn slows the input pulse and moves it to later times.

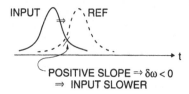

POSITIVE SLOPE $\Rightarrow \delta\omega < 0$
$\Rightarrow$ INPUT SLOWER

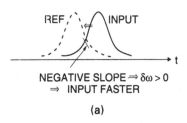

NEGATIVE SLOPE $\Rightarrow \delta\omega > 0$
$\Rightarrow$ INPUT FASTER

(a)

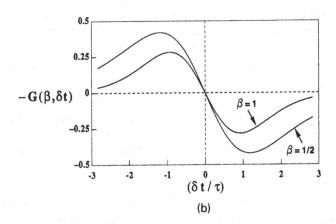

(b)

Fig. 4.3 (a) Intuitive picture of the temporal correcting nature of the nonlinear interaction for $n_2 > 0$ and anomalous group-velocity dispersion in the fiber. (b) Calculated frequency shift of an input pulse because of cross-phase modulation ($\beta = \tau_{input}/\tau_{ref}$).

Formulas for timing correction

We can derive simple formulas to describe the shift of the input pulse center frequency $\Delta\omega_c$ because of cross-phase modulation in the limit of negligible walk-off and dispersion. Standard soliton normalizations (Appendix A) are used in the following equations, where t is local time on the pulse while z is distance along the waveguide. Let us assume that the input u and reference v pulses are given by

$$u(z = 0, t) = \operatorname{sech} t; \qquad |v(z,t)|^2 = A_s^2 \operatorname{sech}^2[\beta(t+\delta t)], \quad (4.8)$$

where $\delta t = t_{\text{input}} - t_{\text{ref}}$ and $\beta = \tau_{\text{input}}/\tau_{\text{ref}}$. The input pulse accumulates a nonlinear phase shift from cross-phase-modulation and is proportional to

$$u(z = L, t) \propto \operatorname{sech} t \exp[i\Delta\Phi(t)]; \qquad (4.9\,(\text{a}))$$

$$\Delta\Phi(t) = \tfrac{2}{3}A_s^2 L \operatorname{sech}^2[\beta(t+\delta t)], \qquad (4.9\,(\text{b}))$$

where L is the length of the semiconductor waveguide. There is also a phase shift arising from self-phase modulation [4.13], but for a symmetric pulse this only leads to a symmetric broadening of the spectrum without a shift in the center frequency.

The shift in the frequency centroid is given by [4.5]

$$\Delta\omega_c = i \int_{-\infty}^{\infty} dt \, u^* \frac{\partial u}{\partial t} \bigg/ \int_{-\infty}^{\infty} dt \, |u|^2 \qquad (4.10)$$

and note from symmetry arguments that $\Delta\omega_c = 0$ when $A_s = 0$. Introducing Eqs. (4.8) into (4.10) we obtain

$$\Delta\omega_c = \tfrac{2}{3}A_s^2 L G(\beta, \delta t), \qquad (4.11\,(\text{a}))$$

where

$$G(\beta, \delta t) = \beta \int_{-\infty}^{\infty} dt \, \operatorname{sech}^2 t \operatorname{sech}^2[\beta(t+\delta t)] \tanh[\beta(t+\delta t)].$$

$$(4.11\,(\text{b}))$$

In Fig. 4.3 (b) we plot the negative of $G(\beta, \delta t)$, which is proportional for anomalous dispersion to $\Delta T/\tau$ of the input pulse, and we find qualitative agreement with the experimental data of Fig. 4.2. The more abrupt drop off of the experimental data may result from laser pulses that are Gaussian rather than hyperbolic secant, while the asymmetry in the experimental data may partially be due to slightly asymmetric laser pulses.

The timing restoration provided by the hybrid TDCS can be tailored by adjusting the characteristics of the reference pulse. The width of the timing window can be adjusted by changing the width of the reference pulse, and the slope of the correction can be adjusted by changing the intensity of the reference pulse. For example, in the calculations of Fig. 4.3 (b), by doubling the pulse width of the reference we increase the timing correction window from 1.8τ ($\beta = 1$) to 2.4τ ($\beta = 0.5$).

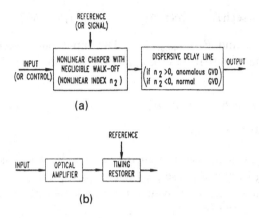

(a)

(b)

Fig. 4.4 (a) Generalized timing restorer consisting of a nonlinear
chirper with negligible walk-off followed by a dispersive delay line.
(b) All-optical regenerator for soliton pulses obtained by combining
the timing restorer with an optical amplifier such as an erbium-
doped fiber amplifier.

Although the experiments are carried out with 415 fs pulses, the
concept can be generalized to different pulse widths as well as to
other nonlinear materials. For example, Fig. 4.4 (a) illustrates a generalized timing restorer. Note that proper operation requires $n_2 > 0$
and anomalous dispersion in the fiber, or vice versa signs, and this
simple picture holds as long as there is negligible walk-off between
the two pulses. Furthermore, we can design an all-optical regenerator
for solitons by cascading a hybrid TDCS with an optical amplifier
such as an erbium-doped fiber amplifier (Fig. 4.4 (b)). The erbium-
doped fiber amplifier can restore the amplitude of the pulses. Because
fundamental solitons try to maintain a constant π-area pulse (cf.
Appendix A), amplification results in both intensity and pulse-shape
restoration for solitons. The remaining time restoration function of a
regenerator can be accomplished by the hybrid TDCS.

4.3 Prechirper to relax timing restrictions in soliton-dragging gates

Now we study the soliton-dragging mechanisms and how the
acceptable timing window for SDLGs can be expanded. We begin by
providing an intuitive picture of cross-phase modulation and walk-off
[4.14] that results in a shift of the frequency centroid ω_c of the control pulse. The formulas from which Fig. 4.5 is derived are given in

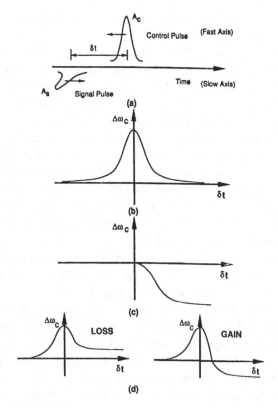

Fig. 4.5 (a) Interaction between orthogonally polarized control and signal pulses. Schematic plot of the shift in the control pulse center frequency $\Delta\omega_c$ versus the pulse separation δt for: (b) complete walk-through of the two pulses; (c) when the two pulses temporally coincide at the input; and, (d) with discrete loss or gain when the two pulses coincide.

Appendix B, where pulse shape changes that may be due to dispersion are neglected. The group velocity dispersion in the fiber translates the frequency shift into a timing shift ΔT.

Consider the interaction between a control pulse (amplitude A_c) along the fast axis and a signal pulse (A_s) along the slow axis (Fig. 4.5(a)). We plot the shift in the control pulse center frequency $\Delta\omega_c$ versus the initial pulse separation δt as the two pulses walk-off. For a complete walk through of two pulses whose shapes remain constant during the interaction the collision is almost completely elastic. As shown in Fig. 4.5(b), the frequency shift curve is symmetric and after the collision the center frequency returns to its original value. There is a slight shift in the time and phase of the control pulse because of the temporary frequency shift; but, in general, the resulting time shift is only a small fraction of a pulse width.

In SDLGs the two pulses start coincident and then walk-off. Thus, each pulse experiences only part of the complete collision and a substantial net frequency shift results (Fig. 4.5(c)). A control pulse along the fast axis slows down and we can obtain a time shift of several pulse widths. Typically the two pulses must be synchronized to within a pulse width to guarantee sufficient time shift from soliton dragging, and this is the origin of the timing restrictions for soliton dragging.

Introducing loss or gain during the walk-off can asymmetrize the interaction. Consider a complete walk through of the pulses when the optical fiber has an abrupt loss or gain at the point where the two pulses maximally overlap. The control pulse still speeds up in the first half and slows down in the second half. However, since the rate of change of the control frequency is proportional to the signal intensity, there will be a net frequency increase (speed up) with loss and a net frequency decrease (slow down) with gain (Fig. 4.5(d)).

From this intuitive picture we see that to relax the timing restrictions we must allow for almost complete walk-through of the two pulses while somehow asymmetrizing the interaction. One further constraint for a cascadable SDLG is that the output control pulse must resemble the input. One way of obtaining asymmetric walk-off without violating the cascadability requirement is to impose distortions on the signal pulse that cause it to change shape during the walk-through [4.15]. For example, we can prechirp the signal pulse with a spectral and temporal change that would occur by propagating in a material with normal group-velocity dispersion. Then, when the signal enters the soliton-dragging fiber, the signal will compress and change in amplitude because of the anomalous dispersion. Consider the configuration for a signal prechirper SDLG as shown in the Fig. 4.6(a). The normal-dispersion prechirper can be an anti-parallel grating pair [4.16], a Gires–Tournoir interferometer [4.17], or a dispersion-shifted fiber with normal dispersion at the operating wavelength. The anti-parallel grating pair can yield large dispersions, but the set-up may be bulky and the gratings will introduce loss. Gires–Tournoir interferometers can be made with low insertion loss and tunable dispersion, but they require multiple passes since the dispersion per pass is small and it may be difficult to obtain a flat dispersion over the entire pulse spectral width. In these experiments we use a normal-dispersion fiber because of the simplicity and

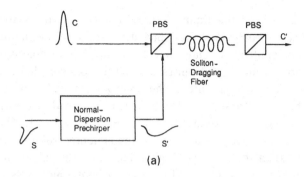

(a)

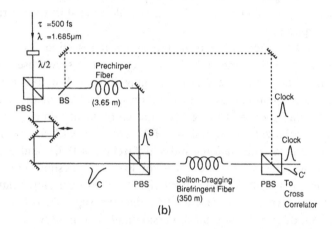

(b)

Fig. 4.6 (a) Schematic of soliton-dragging logic gate with a signal prechirper. The two orthogonally polarized inputs are the control *C* and signal *S*, and the soliton-dragging fiber is surrounded by two polarizing beam splitters (PBS). (b) Experimental apparatus for testing the prechirper concept (BS = beam splitter, $\lambda/2$ = half-wave plate).

compatibility with fiber gates, although we cannot tune the dispersion except by varying the length of fiber or wavelength of the laser.

Normal-dispersion fiber prechirper

We test the prechirper concept in the experimental apparatus of Fig. 4.6(b). Once again, the laser source is a passively mode-locked color-center laser that provides $\tau \sim 500\,\mathrm{fs}$ pulses at $\lambda \sim 1.685\,\mu\mathrm{m}$. The input stage separates the laser beam into a control pulse, a signal pulse and a clock pulse. The signal is passed through a prechirper fiber, and the timing is adjusted so that the control and prechirped signal coincide at the input of the soliton-dragging fiber. A stepper-motor controlled delay stage varies the initial separation between the control and signal pulses δt, and a cross-correlator is used to measure the resulting shift in the control pulse ΔT. The soliton-dragging fiber is 350 m length of birefringent fiber with a polarization dispersion of $\Delta\beta' \simeq 90$ ps/km ($\Delta n = c \times \Delta\beta' \simeq 2.7 \times 10^{-5}$). The control pulse corresponds to an $N = 1$ fundamental soliton with an energy of 50 pJ in the soliton-dragging fiber, and the signal energy is 22.7 pJ in the dragging fiber. Although sufficient time shifts can be obtained with signal energies below 10 pJ, larger energies are used in the experiment to obtain an easily measurable time shift over a range of initial separations.

We use two different fibers in the prechirper section for a comparative study of the time shifts. For a reference case we use 3.65 m of the same fiber as in the soliton-dragging section. We couple in an $N = 1$ soliton signal energy in this section and verify that there is no pulse broadening or narrowing. For the chirp measurement we use 3.65 m of a polarization-maintaining, PANDA fiber with a core diameter of 4.3 μm, a core-cladding index difference of 0.029 and a zero-dispersion wavelength beyond 2 μm. In Fig. 4.7 we show the auto-correlations of the input and output signal pulses emerging from the PANDA prechirper fiber, and we find that the signal broadens by a factor of 3.6 due to the normal dispersion and self-phase modulation.

We plot the shift of the control pulse ($\Delta T/\tau$) versus the initial separation between the two input pulses ($\delta t/\tau$) in Fig. 4.8 for the control pulse along the fast axis. The horizontal axis corresponds to signal delay (i.e., $\delta t > 0$ signal later than control, $\delta t < 0$ signal earlier than control). For the reference case (signal not broadened) the curve is symmetric and we obtain at least a 2τ shift over $-\tau < \delta t < \tau$. With the normal dispersion prechirper fiber the peak shift reduces by a factor of two, but we obtain at least a 2τ shift over $-2.33\tau < \delta t < \tau$. There is a wider window for $\delta t < 0$ because although the control and signal pulses have near complete walk-through, the changing shape of the signal leads to a net chirp. For $\delta t > 1.5\tau$ the pulses hardly overlap and walk-off in opposite directions without interacting. In between the shift ΔT decreases as δt increases because there is less overlap and interaction between the two pulses.

We thus find that by using a 3.65 m prechirper fiber we can broaden the timing window from about 2τ to over 3.3τ. By using

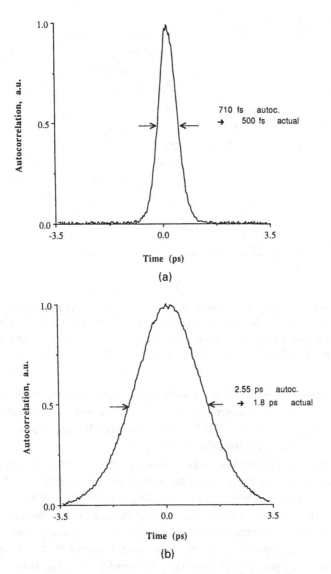

Fig. 4.7 Autocorrelations of the (a) input and (b) output pulses from the 3.65m length of normal-dispersion PANDA prechirper fiber. The signal is broadened by a factor of 3.6 in the prechirper.

longer prechirper fibers and further broadening the signal pulse, we can increase the width of the timing window even more. However, the peak time shift decreases as the window broadens. In fact, for the passive prechirper case it appears that the time shift ΔT integrated

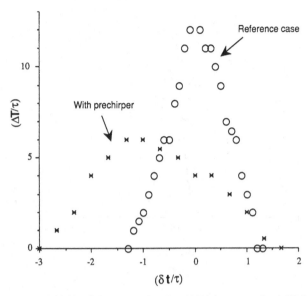

Fig. 4.8 Shift of the control pulse ($\Delta T/\tau$) versus the initial separation between the two input pulses ($\delta t/\tau$) for the control pulse along the fast axis. The reference case corresponds to when the signal is not broadened before the soliton-dragging fiber.

over the range of pulse separations δt may remain roughly a constant. Note that the timing window should only be widened as necessary for reliable operation of the SDLG, since the broader the window the further the signal must shift the control pulse.

Although the passive prechirper arrangement is simple, the penalty for loosening the timing restrictions is paid for by an increase in the minimum switching energy. The switching energy increases because the magnitude of the maximum time shift decreases and because the required amount of time shift is proportional to the acceptable time window. Proper operation of the logic gate requires that the signal energy be adequate to guarantee that the amplitude of the shift ΔT is larger than the maximum time window that will switch the next gate. An alternate means of asymmetrizing the walk-off without necessarily increasing the switching energy is to incorporate an erbium-doped fiber amplifier about one or two walk-off lengths down the fiber. The change in amplitude of the pulses alters the chirp on the two sides of the amplifier and leads to a net chirp even with complete walk-through of the pulses. However, the hardware complexity of the switch increases and the wavelengths are restricted to the gain band of erbium.

4.4 Relaxing timing restrictions using an erbium-doped fiber amplifier

From Fig. 4.5 (d) we see that introducing gain part way during the interaction asymmetrizes the walk-off and, therefore, broadens the timing window for SDLGs. To understand the expected timing broadening in more detail, we numerically solved the coupled nonlinear Schrödinger equation using a split-step Fourier transform technique [4.18]. In the simulations we assume a several walk-off length section of moderately birefringent fiber followed by a discrete amplifier and, then, another longer length of moderately birefringent fiber (see insert in Fig. 4.9 (a)). We are particularly interested in the shape of the timing curve as a function of the initial separation between pulses $\delta t = t_{signal} - t_{control}$. This shape will depend primarily on the position of the amplifier in walk-off lengths, the amplitude of the control pulse in the two sections, and the gain level. The amplitude of the control pulse shift ΔT will depend on the precise values of the birefringence, the signal amplitude, and the length of the fiber. To remain general, we will leave the vertical axis ΔT in arbitrary units; consequently, in the simulations we must only specify the number of walk-off lengths in the first fiber and the gain level. In all cases we will assume that the control pulse is an $N = 1$ fundamental soliton in the final fiber, as required for cascadability of the gate.

In Fig. 4.9 we illustrate the calculated timing curves (shift of the control pulse ΔT versus initial separation between pulses δt) with the same moderately birefringent fiber used on both sides of the amplifier and the control pulse along the fast axis. Curves are plotted for discrete gains of 1.5 and 2 placed at 1, 1.5 and 2 walk-off lengths, l_{wo}, down the fiber (in normalized soliton units $l_{wo} = 1.76/2\delta$, where δ is the normalized birefringence). We see that the timing window can be broadened, although the shape is quite asymmetric. The peak near $\delta t = 0$ arises from the interaction between the control and signal pulses at the beginning of the first fiber. The second feature (at $\delta t < 0$) arises from collision of the pulses in the amplifier, and its magnitude is proportional to the increase in the signal amplitude (i.e., proportional to the gain minus one). The position of this second feature shifts farther from $\delta t = 0$ as the amplifier is moved farther from the input end of the switch.

The ideal case would be to broaden the timing window and have an almost flat-topped, rectangular-like curve for the timing shift. Reference [4.5] describes how from the intuitive chirp picture of Fig. 4.5 (d) we might expect such a curve for an amplifier with 3 dB of

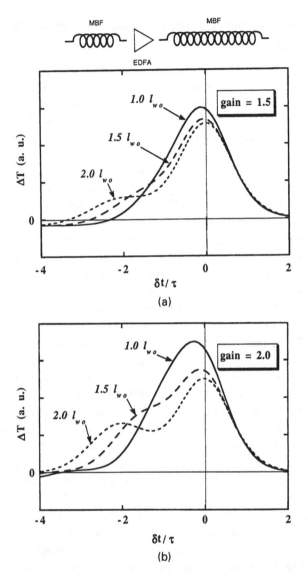

Fig. 4.9 Computed shift of the control pulse ΔT versus separation between control and signal pulses δt when an erbium-doped fiber amplifier (EDFA) is placed 1, 1.5 and 2 walk-off lengths down the fiber. The net gain between the two fibers is (a) 1.5 and (b) 2. The insert shows that the simulations assume two lengths of moderately birefringent fiber (MBF) connected through an EDFA.

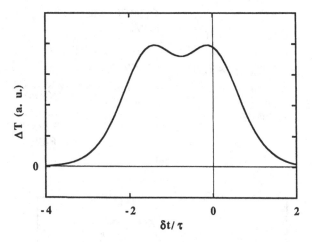

Fig. 4.10 Computed shift of the control pulse ΔT versus separation between the control and signal pulses δt when we neglect pulse shape changes (e.g., $N = 1$ soliton in both fibers) and we have a 3 dB amplifier placed 1.54 walk-off lengths down the fiber.

gain positioned between $1.5l_{wo}$ and $2l_{wo}$ down the fiber; yet, Fig. 4.9(b) does not show such a result. The problem is that we assume identical fiber properties on both sides of the amplifier. If the control pulse is an $N = 1$ soliton in the second length of the fiber, then the amplifier limits the control to $N < 1$ in the first fiber. Consequently, the control and signal pulses broaden while propagating in the first fiber. Since the peak intensity of the pulses decrease by the time the two pulses collide in the amplifier, the strength of the shift also decreases. To achieve the optimal timing curve broadening, we must use different fibers on the two sides of the amplifier; i.e., we need to adjust the fiber core sizes and/or dispersions such that the control pulse is an $N = 1$ soliton in both sections. For example, if we neglect the pulse shape changes associated with propagation [4.5], then we can obtain a nearly flat-topped broadening of the timing window with a 3dB amplifier placed $1.54l_{wo}$ down the fiber (Fig. 4.10).

Experiments in the erbium gain band

To verify the widening of the timing window, we performed experiments using an EDFA. As shown in Fig. 4.11(a), the first section is 25 m of moderately birefringent fiber with a polarization dispersion of $\Delta\beta' = 33$ ps/km ($\Delta n \simeq 10^{-5}$), a zero-dispersion wavelength of 1.5 μm, and a group-velocity dispersion at 1.56 μm of

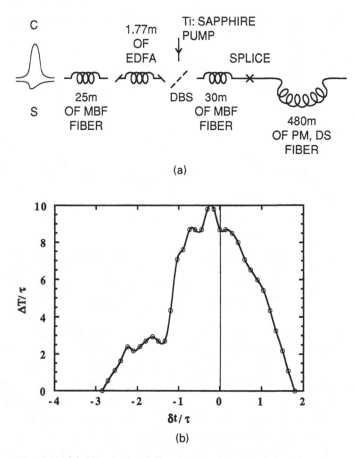

Fig. 4.11 (a) Schematic of fiber lay-out for experimental confirmation of the timing broadening by introducing an erbium-doped fiber amplifier. (b) Experimentally measured timing shift of the control pulse versus initial pulse separation for a net gain of 1.6 and the amplifier about 2 walk-off lengths down the fiber (MBF = moderately birefringent fiber, EDFA = erbium-doped fiber amplifier, DBS = dichroic beam splitter, PM-DS = polarization-maintaining dispersion-shifted). The two ends of the EDFA are beveled to prevent lasing in the amplifier.

$D \simeq 3\,\mathrm{ps/(nm\,km)}$. For $\tau = 0.5\,\mathrm{ps}$ pulses this fiber length corresponds to about $1.6l_{wo}$. The output from this fiber is coupled into a 1.77 m length of polarization-maintaining EDFA with a high doping level of 1660 parts per million of erbium (the ends of the EDFA are beveled to prevent lasing). The core of the EDFA appears asymmetric, so the polarization-maintaining property is due probably to geometric birefringence. The EDFA is pumped at 980 nm by a Ti:Sapphire laser through a dichroic beam splitter that transmits both polarizations at 1.56 μm while reflecting the pump. Finally, the amplified beam is coupled into a 30 m ($\sim 2l_{wo}$) length of the same moderately birefringent fiber, which is fusion spliced to a 480 m length of polarization-maintaining fiber (the splicing loss is $\sim 10\%$). The polarization-maintaining fiber acts as the soliton-dispersive delay line and is a rectangular-shaped fiber that is dispersion-shifted with a zero-dispersion wavelength of 1.415 μm ($D \simeq 5.9$ ps/(nm km) at 1.56 μm). With careful handling of each fiber to maintain the polarization axes, the polarization extinction after all three fibers was better than 10 dB.

We obtain sub-picosecond pulses at $\lambda \sim 1.56$ μm from a passively mode-locked NaCl color-center laser. To modelock within the EDFA gain band, we introduced a wavelength filter consisting of two prisms and a knife-edge [4.19]. The control and signal energies in the first fiber were ~ 12.3 pJ, and the gain in the EDFA was adjusted so that the control energy out of the last fiber was ~ 20 pJ, which corresponds approximately to the $N = 1$ soliton. Therefore, the net gain between the fibers was ~ 1.6 when the coupling and beam splitter losses are included. One serious complication in the experiment is that the EDFA is not a linear amplifier at these power levels [4.20]. For example, the gain for the control pulse decreases in the partially saturated amplifier when the signal pulse is added. Since there is some soliton self-frequency shift in the fiber (the control pulse alone has a spectral shift less than one-sixth of its spectral width), a change in the control pulse amplitude also leads to a time shift (cf. Eqs. (4.4–6)). Since the EDFA has very long lifetimes for the excited state (0.1–1 ms), this is a slow effect and can be compensated for. We also checked the autocorrelation of the pulses emerging from the EDFA and confirmed that there was little pulse shape distortion from dispersion or amplification.

The experimental time shift curve using $\tau = 0.5$ ps pulses from the laser is plotted in Fig. 4.11 (b). We compensate for the amplifier saturation by subtracting out the background shift caused by adding the signal pulse when the control and signal are separated by many pulse widths. Again, since the EDFA lifetime is much longer than the 12 ns separation between pulses from the laser, the shift associated with saturation and SSFS will be independent of the pulse separation. At least qualitatively we see similar behavior as in Fig. 4.9 (a) (gain of 1.5 and first fiber length of $1.5l_{wo}$). Differences in the shape between theory and experiment may be because: (a) the pulses from the laser are not hyperbolic secant pulses as used in the simulations; and (b) the simulations assume a discrete amplifier, while our EDFA has a finite length; and (c) the birefringence in the amplifier causes the pulses to shift their relative positions somewhat (less than a pulse width). ·

Even though the intuitive picture of Fig. 4.5 (d) suggests that an amplifier can broaden the timing window, experimentally it is difficult to obtain a rectangular-shaped timing window. First, we cannot easily obtain $N = 1$ solitons on both sides of the amplifier. Although conceptually it is simple to vary the fiber parameters, in practice it is difficult to obtain two fibers with the desired soliton energies *and* the same birefringence. Second, at the current state of technology, the saturation of the amplifier for our control and signal energies results in undesirable amplitude variations and timing shifts. Further development is required both in decreasing the energy levels in the logic gates and increasing the amplifier saturation intensities.

4.5 Summary

In summary, we have examined the timing problems for a terabit-rate logic system based on soliton-dragging logic gates. For long fiber devices the timing restrictions come from SSFS and environmental effects such as thermal fluctuations. We describe two approaches for relaxing the timing restrictions: (1) reducing the timing jitter on the input bit stream; and (2) increasing the timing window accepted by the logic gates. A timing restorer can be implemented using a nonlinear medium with negligible walk-off followed by a dispersive delay line (e.g., a hybrid TDCS). Furthermore, a soliton regenerator can be made by combining the timing restorer with an optical amplifier such as an erbium-doped fiber amplifier.

To relax the timing constraints for a soliton-dragging logic gate, we must asymmetrize the walk-off between the control and signal pulses. For example, a prechirper at the signal input port increases the timing window at the expense of increased switching energy. An EDFA placed one or two walk-off lengths into the SDLG can also asymmetrize the walk-off. However, in the current state-of-the-art the SDLG energies force the EDFA to operate in a partially saturated regime. One study shows that if logic systems over 100 Gbit/s are to be implemented, then logic gates that operate with net energies close to a picojoule and optical amplifiers with saturation energies around 300 mW will be required [4.1]. Although we have concentrated on SDLGs to be specific, the timing issues and techniques for dealing with them should be generalizable to any terabit rate system.

5

Potential system applications of ultrafast devices

Electronic logic and switching systems may be expected some day to operate at maximum speeds of ~ 20–$50\,\mathrm{GHz}$ [5.1]. Yet, fiber transmission systems already exceed these speeds and are limited only by the electronic demultiplexing speed at the receiver end. One approach to overcoming the electronics bottleneck is to use parallel channels where optics interconnects signals while electronics handles the processing, but there are applications where serial processing is most natural, particularly when the system directly connects to the high-bandwidth fiber systems. Fast serial processing requires ultrafast gates, and the ultimate speed will be found in all-photonic systems where the signals remain as photons throughout the system. Although all the fiber routing and logic devices described in this book are at an early stage of research, it is important to identify and direct efforts toward potential application arenas.

The use of all-optical gates can enhance the system when it is the bandwidth of the switch that limits the performance of the system. Potential applications for ultrafast gates in telecommunications networks include: nodes in local area networks; time-slot interchangers (TSI); header reading in self-routing packet switching; and high-performance front and back ends of long-haul telecommunications fibers (e.g., multiplexers and demultiplexers). These are all examples of time-domain switching functions where the system bottleneck is set by the bandwidth of a few critical components. In addition, computer networks may benefit from high-speed gates for clock synchronization and communication between memory and processors, work station and processors, or multiple-processor machines. Finally, ultrafast gates have been proposed for high-speed encoding and decoding of signals to guarantee secure transmission [5.2].

What significant tasks can be performed with a handful of ultrafast gates? This chapter covers routing switches that can perform demultiplexing at the output of a high-speed transmission system, a logic abstraction of a generalized exclusive-OR (GEO) module with an applications catalog including counters and time-slot interchangers, and the design of a soliton-ring local area network in which digital soliton-dragging and -trapping gates select and decode the header. Whereas the first two are direct applications of the devices described in Chapters 2 and 3, the third is a general architecture for an entire system. A challenge for switching research is to find the applications that provide the compelling reasons needed for the development of this new technology. One big plus for the work described in this book is that fiber gates align the switching effort with ongoing developments in long-haul transmission systems [5.3]. For example, fiber properties are being improved and erbium-doped fiber amplifiers (EDFAs) are being optimized for long-distance soliton transmission, and all of these improvements can be shared by switching systems.

5.1 All-optical demultiplexers

As the bit-rate capacity of long-haul transmission systems continues to increase to 2.5 Gbit/s and beyond, there is a growing need for high-performance front and back ends of fiber systems. The data is transmitted in serial form over a single-mode fiber, and the data may eventually be processed by highly parallel opto-electronic or electronic circuits. However, as shown in Fig. 5.1, there is a nebulous region between these two ends where all-optical fiber switches may play a role in demultiplexing the data. Although the ultrafast gates can perform relatively simple functions, they must process the data serially or convert from serial to parallel format. Whereas bit streams can be time multiplexed using passive, linear components, demultiplexing requires active switches that can handle the bandwidth being off-loaded from the transmission line.

The complexity of the demultiplexing may depend on the format of the data and the control. For example, a self-routing packet switch may be more complicated to implement than a circuit switch. The packets in a self-routing network consist of a header (or address) and payload (or data), and the header must be read to determine the destination of the packet. In a circuit switch a path is set up, and then the data flows along this fixed path from source to destination. Also, the bandwidth of the demultiplexer will depend on whether the routing follows a fixed, periodic pattern or whether the routing is data dependent. If the data is spatially separated using a perfectly periodic

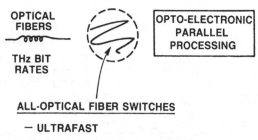

OPTICAL
FIBERS
~~~~~~
THz BIT
RATES

OPTO-ELECTRONIC
PARALLEL
PROCESSING

ALL-OPTICAL FIBER SWITCHES

 — **ULTRAFAST**

 — **SIMPLE FUNCTIONALITIES**

 — **SERIAL PROCESSING**

Fig. 5.1 All-optical fiber switches may be used as demultiplexers or processors between optical fibers and electronic and opto-electronic parallel processors.

function, then the bit-rate can be halved at each stage and the demultiplexer may be driven sinusoidally. Consequently, the bandwidth of the demultiplexer needs to be a narrow band centered around a subharmonic of the pulse-repetition rate. As an example, lithium-niobate switches with phase-matched electrodes have been used to demultiplex 72 Gbit/s data down to 36 Gbit/s [5.4]. The periodic separation of data does not require reading the data, but instead involves swinging a light-pipe back and forth at a regular rate. All-optical switches are bit-rate switches (bandwidth from DC up to the bit rate) that can provide arbitrary demultiplexed patterns or data-dependent routing. The analogy in electronics would be the difference between a ring oscillator and a logic circuit: whereas a device may be used in a 200 GHz ring oscillator, it cannot necessarily perform arbitrary logic operating at 200 Gbit/s.

In addition to work done on the implementations of demultiplexers or AND-gates (Chapter 2), considerable effort has continued in Kerr effect, four-wave mixing and nonlinear optical loop mirror (NOLM) devices, and their bit rates and performance have improved. Current experiments in demultiplexers are targeted toward several key objectives:

(1) operating beyond 2.5 Gbit/s, which is the speed for which next-generation long-haul systems are being designed, and even beyond 50 Gbit/s, which should be beyond electronic speeds;

(2) lowering the switching energy approaching 1 pJ/bit, since a terabit-rate of data at a picojoule per bit will require lasers with a watt of average power;

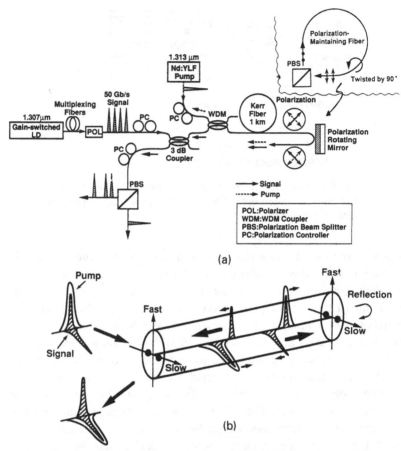

(a)

(b)

Fig. 5.2 (a) Experimental set-up for dual-path optical demultiplexer based on the Kerr effect. (b) Principle of birefringence compensation in a Kerr medium [5.5].

(3) achieving arbitrary demultiplexing of bit streams, which would warrant the use of bit-rate devices.

Three recent experiments show approaches to each of these challenges.

To illustrate the high bit rates achievable by all-optical techniques, Morioka, et al. [5.5], have demultiplexed a 50 Gbit/s data stream using the Kerr effect in a 1 km long PANDA fiber. As described in Section 2.1, Kerr modulators use the intensity-induced change in polarization state followed by an analyzer to perform a routing function. The experimental apparatus for the dual-path Kerr demultiplexer is illustrated in Fig. 5.2. The 10 ps pulse-width signal pulses are derived from a 1.307 μm gain-switched semiconductor laser diode

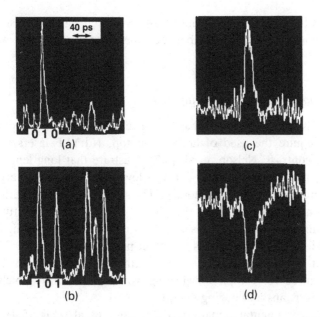

Fig. 5.3 Experimental results from Kerr demultiplexer: (a), (b) 50 Gbit/s pulse operation; (c), (d) Kerr modulation profiles for cw operation [5.5].

(LD) and then multiplexed up to 50 Gbit/s, using a series of 3 dB couplers to interleave the data. The pump pulses are from a 1.313 μm mode-locked Nd:YLF laser whose output is compressed to 10–15 ps with a repetition rate of 82 MHz.

The two pulse streams are combined in the 1 km long Kerr fiber with the pump along one polarization axis and the signal at 45° to the axes. At the end of the fiber, the polarization is flipped using a fiber-pigtailed polarization beam splitter, whose two output ports are connected to each other so that the polarization of the returning pulse is flipped to the crossed state (see Fig. 5.2(b)). To obtain stable demultiplexing at these high repetition rates, Morioka, et al., borrow an idea from the nonlinear optical loop mirrors [5.6]: namely, to establish two identical paths, simply propagate the signal in two directions along the same fiber. Therefore, to compensate exactly for the birefringence, the control and signal pulses travel to the end of the fiber, are flipped to the orthogonal polarization, and then propagate back through the same fiber (Fig. 5.2(b)). The switching results are shown in Fig. 5.3, where (a) and (b) correspond to the switched and unswitched parts at a switching power of 1.5 W. Optical Kerr modulation profiles were also observed using a cw-signal as shown in Figs.

5.3 (c) and (d), and the 20 ps width of the switched signal corresponds to the round-trip group delay between the control and signal wavelengths.

Loop mirror as a demultiplexer

This experiment emphasizes speed, and the high powers involved require the used of large, table-top, Nd:YLF lasers as the pump. By contrast, Nelson, et al. [5.7], illustrate that long lengths of fiber in an NOLM (see Section 2.4) can lower the switching energy to levels approaching a few picojoules. The target of their experiment was to demultiplex signals in the gigabit regime at sufficiently low energies that semiconductor lasers (SCL) could be used throughout. They used an NOLM with 6.4 km of dispersion-shifted fiber. If the interaction length equals the entire length of this device, then the power required for complete switching would be 160 mW, which for 10 ps pulses means a switching energy of 1.6 pJ.

In the experimental apparatus of Nelson, et al. (Fig. 5.4), the pump pulses were from a 1.53 μm gain-switched SCL driven at 2.5 GHz and were compressed to ~16 ps in a 700 m length of fiber with anomalous dispersion (soliton regime). These pulses were then amplified in an EDFA to average power levels of 20 mW. The data or probe source was a mode-locked SCL operating at 1.56 μm with a repetition rate of 10 GHz, which was then interleaved in a fiber Mach–Zehnder device to produce a 20 GHz pulse train. The coupler in the loop had a 50:50 ratio at the data 1.56 μm wavelength and a

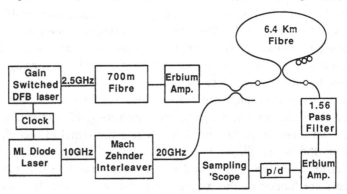

Fig. 5.4 Schematic of the experimental apparatus for testing a 6.4 km long nonlinear optical loop mirror demultiplexer (DFB = distributed feedback, ML = mode-locked, p/d = photodiode, Amp = amplifier) [5.7].

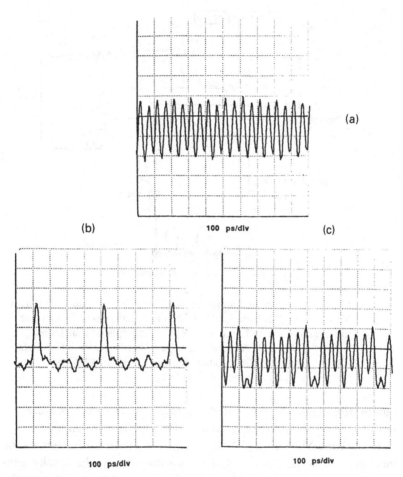

Fig. 5.5 Experimental results from the 6.4 km long nonlinear optical loop mirror. (a) Mode-locked 1.56 μm pulse train at 20 GHz repetition rate. (b) Switched output with the loop in "reflecting mode" – i.e., only every eighth signal that coincides with the control pulse is transmitted. (c) Switched output with the loop in the "transmitting mode" – i.e., only pulses that do not coincide with the control are transmitted [5.7].

100:0 ratio for the 1.53 μm pump. The NOLM can be adjusted for operation in a reflection or transmission mode by adjusting the state of polarization within the loop through a polarization controller [5.8]. In Fig. 5.5 we show the experimental results, where a 1.56 μm pass filter is used before the detector to block the pump. Figure 5.5 (a) corresponds to the 20 GHz signal train, and (b) are the switched pulses while (c) are the complementary unswitched pulses. Every eighth pulse is switched

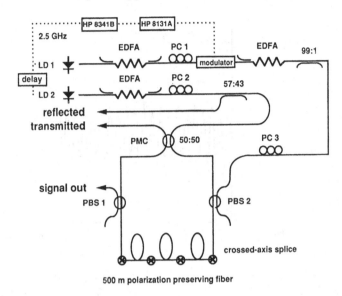

Fig. 5.6 Experimental apparatus for testing a nonlinear loop mirror with orthogonally polarized control and signal pulses. To increase the interaction length, the fiber is cross-axis spliced four times. (LD = laser diode; EDFA = erbium-doped fiber amplifier; PC = polarization controller; PBS = polarizing beam splitter, PMC = polarization-maintaining coupler) [5.9].

out by the demultiplexer, but the contrast is imperfect because the power is lower than the required level for complete switching.

In both the Morioka and Nelson implementations, a fraction of the incoming bit stream is directed to the output and the locally generated pump pulses are blocked. Therefore, the output may be degraded because of loss, dispersion, cross-talk or timing jitter as the data traverse the system. Whitaker, et al. [5.9], describe another NOLM demultiplexer design using orthogonally polarized pulses that is regenerative in the sense that the signal pulses are replaced by new control pulses from the power supply. An advantage of regenerative switches is that they provide some insensitivity to timing jitter at the input; i.e., the output pulse power level, timing, and pulse shape are determined by the local laser power supply rather than the incoming pulses. Whitaker, et al.'s, design is a combination of the Moores, et al. [5.10], and Avramopoulos, et al. [5.11], experiments that were described in Section 2.4. They demonstrate complete switching of an arbitrary pattern of pulses from a 2.5 Gbit/s pulse train with tolerance to timing errors in the signal stream of up to 350 ps.

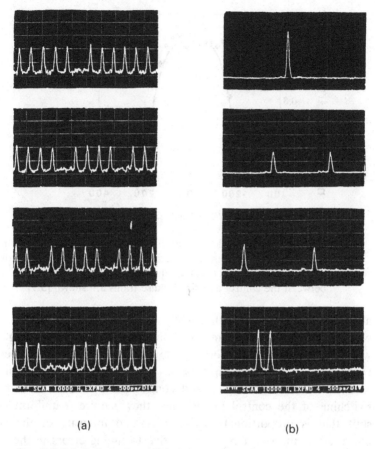

Fig. 5.7 (a) Reflected and (b) the complementary transmitted outputs from the nonlinear optical loop mirror. Arbitrary pulse patterns can be switched based on the encoding of the signal bit stream. A narrow band interference filter is used to block the leakage from the signal laser diode in transmission and reflection [5.9].

In their experimental apparatus of Fig. 5.6, the signal and control pulses were of 80 ps duration and were derived from gain-switched laser diodes (LD) and amplified in EDFAs (the wavelength of LD1 was 1.534 μm and of LD2 was 1.531 μm). The signal pulses were encoded using a lithium-niobate modulator and then further amplified in an EDFA to an average power of > 80 mW. The control pulses propagate in both directions around the loop, while the signal pulses are injected through a polarizing beam splitter, travel in only one direction and then are removed by another polarizing beam splitter. The various polarization controllers are adjusted so that the control

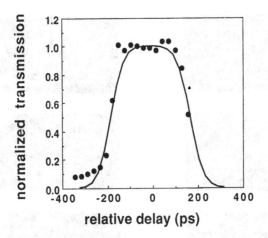

Fig. 5.8 Timing window or jitter tolerance of the regenerative non-linear optical loop mirror. Experimental results (dots) and theoretical prediction (solid curve) for the transmission versus relative delay between the control and signal pulse trains are shown [5.9].

and signal pulses are orthogonally polarized in the NOLM. The NOLM consists of a 500 m length of polarization-maintaining fiber with four cross-splices, where the slow and fast axes of the fiber are exchanged to effect multiple passes. As described in Section 2.4, walk-off between the signal and control pulses results in complete switching of the control pulses since they acquire a uniform phase shift that is proportional to the integrated intensity of the signal pulse. Also, the switching is insensitive to timing errors in the signal stream if the time window is longer than the pulse duration.

Arbitrary pulse patterns switched from the NOLM are shown in Fig. 5.7, where (a) corresponds to the reflected and (b) the complementary transmitted pulse patterns. Imperfect polarization extinction in the polarization beam splitters as well as imperfect alignment of the polarization axes of the fiber during splicing necessitated the use of a narrow band interference filter to block the signal leakage in transmission and reflection. The aggregate signal-pulse repetition rate was 60 MHz, and the switching energy was ~1.3 nJ, which corresponds to a peak power of ~17 W. To test the width of the timing window, the relative delay between the electrical signals driving the two laser diodes was varied and the transmission was monitored. Figure 5.8 shows the experimental results (dots) and the theoretical predictions for squared hyperbolic secant pulses of 80 ps duration and a birefringence slip of 350 ps (solid curve). The transmission changes less than 6% for a timing window of 250 ps, and the half transmission width is 350 ps. This insensitivity to the timing jitter within a finite window may also be used to correct for timing errors in long-haul transmission systems [5.12].

5.2 GEO-modules and all-optical processing

The previous section concentrated on the application of routing switches to high-performance ends for a telecommunications fiber system. Let us change gear now and consider design issues for all-optical computing and logic systems that are based on soliton-dragging logic gates. A generalized exclusive-OR (GEO) module is a Boolean and connectivity-complete logical representation of the soliton-dragging logic gate [5.13]. For designing systems and architectures we use GEO-modules as a tool to abstract from the physical implementation of the soliton logic gates. One key advantage of this approach is that the system architecture is independent of the device hardware, allowing devices to be upgraded provided their functionality remains the same.

As new device concepts are introduced, system designers need to consider what is the optimal building block for all-photonic systems. A natural choice for the fundamental building block is a Boolean-complete module that reflects or parallels directly the physics of the devices. For electronic systems, the three-terminal NOR-gate may be the obvious choice. On the other hand, for systems based on lithium-niobate switches, the natural choice is five-terminal directional couplers [5.14]. For fiber-based logic systems, the natural basis is a GEO-module.

This section includes an applications catalog for a handful of GEO-modules, showing the utility of the modules for implementing some fundamental computational blocks using a minimum number of components. First, the GEO-module is introduced, with an explanation of its use for logic operations and a description of combinatorial circuits using two GEO-modules. Then, a single GEO-module sequential circuit is presented, along with relevant issues of latency and minimal-component design. A counter is given as an example of a non-trivial finite-state machine, and a time-slot interchanger is introduced to provide communication between time-multiplexed processors. These examples point out that if device designers find that the physics of a device resembles a GEO-module, then mimicing NOR-gate or directional-coupler circuits is uneconomical in terms of gate count. The minimal component implementation will be in terms of the natural building block that is Boolean complete.

GEO-modules and fiber gates

As described in Section 3.1, ultrafast, cascadable, all-optical soliton-dragging logic gates have been demonstrated in optical fibers connected through and followed by polarizing beam splitters. The logic operations are based on time shifts from soliton dragging in a clocked, digital optical processor. Thus, the fundamental building block in these gates consists of a birefringent fiber surrounded by two polarizing beam splitters (PBS), as shown in Fig. 5.9(a) An equivalent circuit for the logic operation consists of an exclusive-OR (XOR) gate and two AND-gates (Fig. 5.9(b)). The action of soliton dragging corresponds to the XOR operation, and the role of the output PBS to look along one or the other axis is represented by two AND-gates. More generally, we can replace the fiber module by a four-port GEO-module with inputs a and b and outputs a' and b', where $a' = a \cdot \bar{b}$ and $b' = \bar{a} \cdot b$. In other words, the input on one axis travels to the output unless annihilated by an input on the orthogonal axis.

A single GEO-module can implement an inverter or an XOR-gate. As shown in Fig. 5.10(a), an inverter consists of a clock (CLK) or power supply "1" along one axis and a signal along the other. Since the two arms of the GEO output are orthogonally polarized, both outputs can be combined using a polarization combiner that may be a polarizer or PBS placed at 45° relative to the principal axes of the fiber. An XOR-gate consists of a GEO-module followed by a polarization combiner, as illustrated in Fig. 5.10(b).

Two GEO-modules can be interconnected to provide high functionality. Figure 5.11(a) shows that by using each GEO-module as an inverter, we can implement a broadcast or buffer switch (assuming the power supply is larger than the signal) to provide multiple copies of the input. Since the signal is replaced by a new CLK pulse, this may also serve as a pulse regenerator or a clock-ANDer to restore the logic level. The NOR-gate described in Section 3.1 is shown in Fig. 5.11(b), where along the CLK axis we find A NOR B and along the orthogonal axis we find A AND B. This NOR-gate is analogous to an electronic relay where the power supply voltage connects to the output unless a signal closes the contact and shunts the voltage. Note that A and B act in series and the same idea can be extended to an N-input NOR-gate. Since a NOR-gate is Boolean complete, GEO-modules must also be Boolean complete. Furthermore, for the same hardware used in the NOR-gate, we can also implement a routing switch (Fig. 5.11(c)). Depending on the value of the control C, signal

(a) Fiber Module

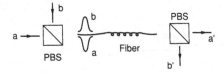

(b) Circuit Representation

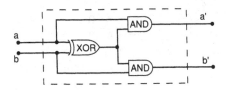

(c) Generalized Exclusive-OR (GEO) Module

Fig. 5.9 (a) Basic building block for a soliton-dragging logic gate consisting of a birefringent fiber surrounded by two polarizing beam splitters (PBS). (b) An equivalent circuit representation of the logic function performed. (c) A generalized exclusive-OR (GEO) module in which the input along one axis propagates to the output unless blocked by an input on the orthogonal axis.

(a) Inverter

(b) XOR - Gate

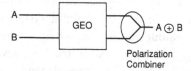

Fig. 5.10 A single GEO-module can implement (a) an inverter or (b) an XOR-gate.

(a) Broadcast or Buffer

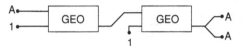

(b) NOR - Gate and AND - Gate

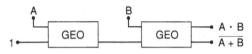

(c) Routing Switch

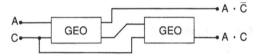

Fig. 5.11 Two GEO-modules can be configured as (a) broadcast or buffer switches, (b) NOR- and AND-gates, and (c) routing switches.

A will appear at one of two separate ports. Note that this is a logic switch (the logic value of *A* is routed) rather than a physical switch (i.e., in a physical switch the photons corresponding to signal *A* would be routed). A directional coupler or exchange/bypass switch can be implemented with two such routing switches and a pair of 3 dB couplers.

Since the GEO-module is a passive element, obtaining gain requires a CLK input with amplitude greater than the signal input. In addition, the output of interest must be along the axis corresponding to the CLK. For example, the NOR-gate in Fig. 5.11 (b) has a CLK input and the output is along the CLK axis; the gate has been demonstrated to have a gain or fan-out from six [5.15] to thirty [5.16]. On the other hand, if we use the same configuration as an AND-gate, there is no gain since the output is just signal *B*. Likewise, the inverter in Fig. 5.10 (a) can have gain, but the XOR-gate in Fig. 5.10 (b) will not have gain. We can still build a system out of gates without gain, but then periodically in the system we need to insert clock-ANDers as in Fig. 5.11 (a) to restore the logic level and timing. How often the clock-ANDing is necessary depends on how rapidly the signal deteriorates in the system.

Single GEO-module sequential circuit

In addition to demonstrating the simplest case of combinatorial circuits using two GEO-modules, we can also make sequential circuits by adding feedback and memory. A fiber delay line acts as a shift register in an all-photonic system since photons must travel at the speed of light in the medium. We require that propagation in the delay line does not distort the pulses, which can be satisfied in short fiber lengths or by using solitons in longer fibers [5.17]. When we loop back on a guided-wave device whose transit time is greater than the interpulse separation, we must explicitly account for the latency: typically the two data "bits" or packets to be interacted must be spaced at least by a latency interval. The latency can be hidden architecturally by properly pipelining processes [5.14, 5.18]; for example, different packets or users can be time interleaved and processed sequentially.

The simplest case of a single GEO-module sequential circuit is illustrated in Fig. 5.12. When the input corresponds to a string of ones, then the circuit acts as a ring oscillator or a multivibrator, as was shown experimentally in Section 3.1 for fiber soliton-dragging logic gates. To perform some processing, we assume that the input is encoded as a sequence of control packets C_i, signal packets S_i and reset packets R_i. The packets are either a gate-latency long or they are time multiplexed with other processes to be at least a latency interval apart. At the output, we obtain $S_i \text{ AND } C_i$ or $S_i \text{ NOR } C_i$ by looking at the proper port at the correct time. Therefore, this circuit

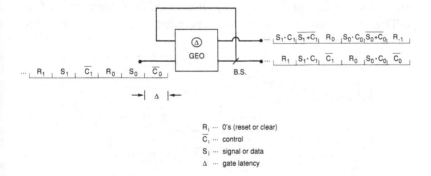

Fig. 5.12 The simplest single GEO-module sequential circuit. The input can be encoded as a sequence of control packets C_i, signal packets S_i, and reset packets R_i. A beam splitter (BS) divides the output beam into two parts, one of which is fed back to the input.

can be used to perform logic operations or be self-routing in a demultiplexer. The reset packet consists of 0s and is used to clear the feedback loop before the next process. This example illustrates a general principle that as we try to minimize the number of components, more encoding and decoding is required at the input and output. Furthermore, more time slots are required to perform an operation, which may not be a drawback if high-speed gates are used.

Serial counter as a finite-state machine

The next important element is a processor with memory and several states: a finite-state machine. Jordon [5.18] suggests that a simple, yet significant, example of a finite-state machine is a four-bit, scale-of-16 serial counter, as sketched in Fig. 5.13. The half-adder has two inputs (X and Y), and outputs carry ($C = X$ AND Y) and sum ($S = X$ XOR Y). On the left is a four-bit increment signal consisting of a one in the low-order bit position and three zeros. Below the half adder is a stored four-bit count value, with low-order bit at the left. Above the half adder is the carry-bit loop. Under properly operating conditions (i.e., reset with input of one only after counting up to sixteen), the OR-gate has at most a single one entering at any given time and, therefore, the OR-gate can be implemented with a beam splitter, a 3 dB coupler or any form of a wired-OR.

Two GEO-modules can be interconnected for a minimal-component implementation of the half-adder (Fig. 5.14(a)). As shown in Fig. 5.10(b), an XOR-gate for the sum can be made with a single gate. Although Fig. 5.11(b) shows one implementation of an AND-gate, it is preferable to implement the carry using the

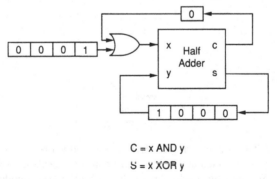

C = x AND y

S = x XOR y

Fig. 5.13 Block diagram of a bit-serial counter. The shift registers are just fiber delay lines of the appropriate length.

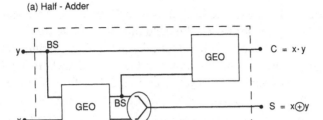

(a) Half - Adder

(b) Bit Serial Counter

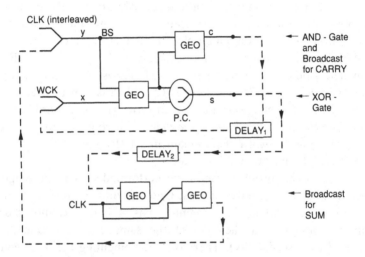

Fig. 5.14 (a) Minimal component implementation of a half-adder using two GEO-modules. (b) Four GEO-module design for a scale-of-16 bit-serial counter.

configuration in Fig. 5.11 (c) for reasons that will be clear below. Note that the first GEO-module in the sum and carry have the same input and, therefore, can be shared in generating both outputs. Since all inputs in Fig. 5.14 (a) are used efficiently and no CLK inputs are present, one drawback of this half-adder is that it does not restore the logic level of the outputs.

To insure proper operation of the counter, we must provide clock-ANDing for the half-adder outputs. Notice that after an appropriate delay C is fed back through the same path. Moreover, the AND circuit of Fig. 5.11 (c) is identical to the broadcast gate of Fig. 5.11 (a) with the control input replaced by a CLK input. Consequently, if we use two clock cycles and interleave CLK pulses with Y, then the AND-gate can act as the clock-ANDer for C. We need only to use one further broadcast switch to clock-AND with S. Our minimal component implementation of the bit serial counter is shown in Fig. 5.14(b), where WCK is a one pulse every fourth clock cycle (see Fig. 5.13): to divide the clock frequency by four may require an additional two GEO-modules or routing switches. As discussed earlier, the two shift registers can be implemented with proper lengths of fiber (represented by "delay").

All-optical time-slot interchanger

As we use time multiplexing to achieve parallelism in time, we must also allow for communication between data streams. Communication between time-multiplexed data streams can be accomplished by use of a time-slot interchanger (TSI). Just as a NOR-gate provided Boolean completeness, a TSI allows for connectivity (in time) completeness. The simplest TSI problem is that two data units, A and B, are traveling down a fiber in series (not necessarily in adjacent time slots). Depending on the value of a control signal, the ordering of A and B will either remain intact (BAR state) or they will interchange time slots (CROSS state).

Note that this problem cannot be performed in a single length of fiber that has two orthogonal axes. One axis must be reserved for the control signal, and in the remaining axis A and B cannot slide in opposite directions if they are at the same central frequency (as required by cascadability). Therefore, implementing the TSI requires at least two GEO-modules. One method for changing time-slot sequencing is to store the data in random-access memory, and then read the data out in the desired order. Since present memory-access times are much slower than the GEO-module switching times, this method would limit the overall operating speed of such a device. Another design that incorporates a feedback loop to reorder time slots requires that the time slots to be exchanged must be separated by more than a gate latency [5.18, 5.19]. This restriction is prohibitive for ultrafast processing when gates with large latency are used.

However, by using a feed-forward architecture that eliminates the need for a discrete optical memory or feedback, we design a TSI that can reorder adjacent time slots and that is not limited by the latency of the gate. The principle behind the feed-forward designs is to use an input stage to generate several copies of the time-multiplexed input sequence. The copies are then time delayed and recombined to generate two signals: one with the original ordering, the other with the ordering interchanged. Each signal is then individually processed by a control circuit, after which the signals are combined through a wired-OR at the output. Depending on the control signal, the input will arrive at the output in either the BAR or CROSS state. There-fore, the random-access memory of an electronic TSI is replaced by time-differencing techniques.

One design for the feed-forward TSI requires three GEO-modules and one common control; another requires two GEO-modules each with independent control; a third requires three GEO-modules each with independent control. The input time-multiplexed sequence con-sists of time slots temporally spaced by δ units. In the first design, three copies of the input are made and two are passively time delayed by δ and 2δ units. The undelayed and 2δ delayed copies are then combined using a wired-OR gate as shown in Fig. 5.15 (a). This inter-changes the order of the time slots A and B in the center of signal S_1. Note that the wired-OR gates can be used to combine the signals, provided we guarantee that two ones do not arrive simultaneously and interfere at the wire-OR: this requires that two time slots be left blank on either side of the time-slot pair being operated on. A three-GEO-module selection switch then picks one of two bit streams (Fig. 5.15 (b)), as determined by the control signal (CONTROL) that is coded to look only at the bits of interest. Because of the various wired-OR's, an amplifier is required at the output to restore signal levels for cascadability. Also, an amplifier will probably be required at the input to compensate for splitting losses.

The second design is a minimal-component, feed-forward TSI. The price of reducing the number of GEO-modules from three to two is an increase in complexity of the control signals. This design has an input stage identical to that of the previous design and provides sig-nals S_1 and S_2. As shown in Fig. 5.16, each signal is fed into a separate GEO-module with independent control sequences, which are coded to look at the time slots of interest. At the output a wired-OR is used to combine the signals, and amplifiers restore the signal level. One drawback with both of these designs is that the two-blank time-slot restriction reduces system throughput by fifty percent for pipe-lined operations. In addition, simplicity of system design scales poorly with increasing numbers of time slots to be interchanged. For exam-ple, more time slots can be interchanged by cascading several two-packet TSIs [5.19], but substantial bookkeeping is required then to keep track of the blank packet positions.

(a) Input Stage

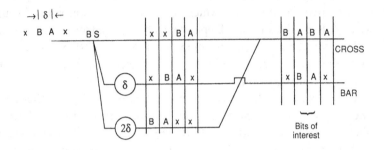

(b) Selection Switch

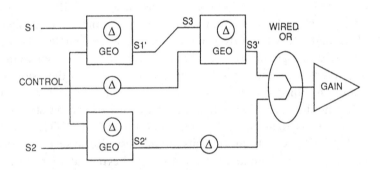

CROSS STATE   CONTROL = 0 1 1 0

BAR STATE   CONTROL = 0 0 0 0

Fig. 5.15 One design of a feed-forward time-slot interchanger.
(a) The input stage makes three copies of the input and time displaces them. (b) Three GEO-modules are used as a selection switch
to pick between the CROSS and BAR states.

By including one extra GEO-module with an independent control
to the previous design, we can design a completely feed-forward TSI
without any blank bit restrictions (Fig. 5.17). As with both previous
designs, the input time-multiplexed sequence must be copied in tripli-
cate and two copies delayed by δ and 2δ units. However, unlike the
previous designs, each signal is processed in parallel using a separate
GEO-module with independent control that is appropriately coded to
look at the time-slot pair of interest. Since blank time-slots are no

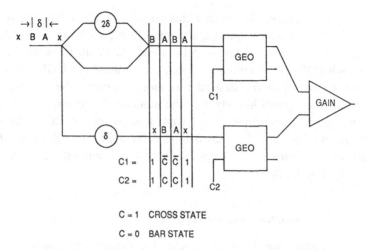

Fig. 5.16 Minimal component design of feed-forward TSI. The coding required for the control sequences is also shown.

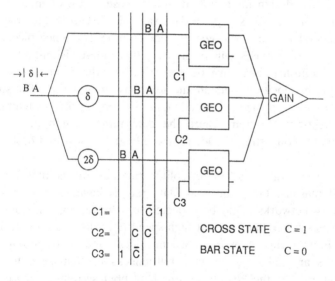

Fig. 5.17 Feed-forward TSI without blank time-slot restrictions. The coding required for the control sequences is included.

longer required, the control sequence must operate at the same bit rate as the input time-multiplexed signals.

In general, in this third design we can switch an arbitrary number of time slots in a feed-forward fashion with linear hardware complexity. For example, suppose that the data frame consists of N packets

$a_N, a_{N-1}, \ldots, a_2, a_1$. To reorder this set, we use an N-TSI, where Fig. 5.17 is generalized by splitting the input into $2N-1$ lines, with delays sequencing from no delay, δ delay units, 2δ delay units all the way through $(2N-2)\delta$ delay units. Then, $2N-1$ number of GEO-modules are used to select the desired time-slot ordering, which means that the hardware grows linearly with the number of packets. Generalizing from Fig. 5.17, there are $2N-1$ control sequences, each of which are N packets long, and C_j starts δ time delay after C_{j-1}. Because of the increased splitting losses, amplifiers with large gain will be required at the input and output.

5.3 A soliton ring network

Examining a system architecture helps us to understand the capabilities that ultrafast logic gates, such as those described in Chapter 3, can bring to telecommunications networks. The example below is the design for a slotted, soliton-based local or metropolitan area network that has a peak data rate of 100 Gbit/s and that may span several tens or hundreds of kilometers [5.20]. Since picosecond or femtosecond pulses will be used in this time-division-multiplexed system, soliton pulses must be used to avoid the deleterious effects from group-velocity dispersion and self-phase modulation. The soliton ring network uses digitally encoded, self-routing packet switching and is a "light-pipe" system where the data remains in optical format throughout, converting to electronics only at the host and destination nodes.

This design of the 100 Gbit/s soliton ring network is interesting for several reasons. First, the peak data rate is beyond the limits that electronic networks might be expected to achieve; e.g., this design is 1000 times faster than current high-speed fiber-distributed data interface (FDDI) ring networks. Second, local and metropolitan area networks are the next expected bottlenecks in high-speed telecommunications. As the bit-rate in long-haul fiber systems increases, the bit-rate bottleneck shifts to the systems that interface to the long-distance data highways. Third, ring networks are particularly limited by the switch bandwidth at the nodes because the average bit-rate available to each user is the bandwidth of the access nodes divided by the number of users on the network. Therefore, increasing the switch bandwidth increases the overall capability of the network. This design is a pipelined, feed-forward architecture that exploits the bandwidth while downplaying the latency of the ultrafast logic gates. Also, a ring

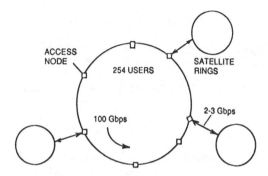

Fig. 5.18 System overview of the soliton-slotted ring network.

network is one potential application of all-optical gates that requires only a handful of gates per node. The purpose behind this design is to provide a complete system architecture that uses the features of ultrafast gates in order to discover and address the relevant issues regarding their use in systems.

This study finds that ultrafast soliton-dragging and -trapping logic gates can enhance the capabilities of a soliton-ring network by processing the packet header at the bit rate. A node in a local or metropolitan area network may be thought of as a post office with a conveyor belt of packages entering. The speed of the conveyor belt and the number of packages that enter are limited by the speed with which a postal worker can read the addresses and sort the packages. If the worker is a speed reader, then the package flow rate can be greatly enhanced. In this network design, four logic modules are used at a node to read the address of a data packet and to decide if the packet destination corresponds to this node. Therefore, the ultrafast soliton gates serve as the speed-reading postal worker.

Figure 5.18 depicts the hierarchical ring network, with a slotted ring having up to 254 user-access nodes and a peak bit rate of 100 Gbit/s For an efficient design where the network size is not dominated by the nodes, each node should be spaced by several gate latencies. Each backbone user can support an average of up to a few Gbit/s out of its host, and the nodes could themselves be a connection to other satellite rings. These satellite rings could then run at about 10 Gbit/s, and might be implemented in more conventional technologies. In this straw-man example we select a trivial protocol in which the header is examined to see if the packet is empty or if the packet has arrived at its destination. If the header matches the local address (*ForMe* true), then we remove the packet and replace it with

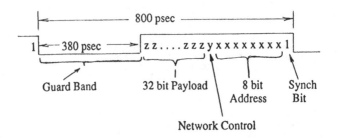

Fig. 5.19 Data representation in the soliton ring network. The header and payload share the same digital representation. The synch bit is always "1" for clock synchronization, and there is an 8-bit address for each node. One network-management bit is used for failure safeguards and a scavenger node, and the data corresponds to a 32-bit payload. To enable synchronization and packet switching, there is a time guard band of 380 ps between the packets.

an empty packet. On the other hand, if the packet is empty (*Empty* true) and another packet is queued in the node buffer, then we replace the empty packet with the new packet.

The self-routing packet (Fig. 5.19) is a 42-bit quantity that has 10 bits of header and occupies 420 ps. In the header, eight of the bits correspond to the address, one bit is used for synchronization, and the last bit is used to enable packet cleanup. An empty packet has every address bit equal to "1" to make empty-packet detection as easy as possible and to maintain synchronization. There are, thus, 254 user addresses available in the 10-bit header after accounting for the null and synchronization bits. The data portion of each backbone packet is a 32-bit computer word, so many electronic end users can communicate to the network with the same flexibility as with their internal backplanes. Also, to enable synchronization and packet switching, there is a time-guard band of 380 ps between packets, so the packet period is 800 ps and the packet rate is 1.25 GHz. We, therefore, trade-off system bandwidth by inserting the guard bands, which is the least expensive system resource, to solve the severe problems of synchronization and packet switching with opto-electronic switches.

The design focuses on the network access nodes (Fig. 5.20) in which the ultrafast gates are used to decode the header, which has the same physical format as the data. The code-matching logic module operates at the bit-rate and checks if the header matches the local address or corresponds to an empty packet The output from the code-matching circuit are electronic signals that control a network of

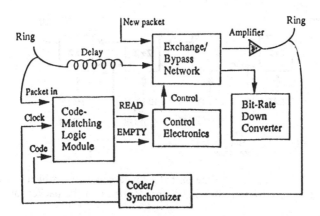

Fig. 5.20 Block diagram of the network access node. The code-matching logic module operates at the bit-rate, while the exchange/bypass network reconfigures at the packet rate. The delay in the upper arm compensates for the latency through the code-matching logic module, and the local clock is synchronized with the ring data.

routing switches. The delay in the upper arm compensates for the latency in the logic module. The exchange/bypass network routes the incoming packet or a new packet, operates at the packet rate, and can be reconfigured in the time-guard band between packets. When the packet reaches its destination, it is demultiplexed or bit-rate down-converted to speeds accessible by electronic shift registers [5.21]. An optical phase-lock loop [5.22] is used to synchronize the local clock to the ring data, and erbium-doped fiber amplifiers are used to compensate for the insertion and splitter losses.

The 2×2 switches in the exchange/bypass network of Fig. 5.21 route the packet depending on the signals from the code-matching logic module. These switches operate at the packet rate and can be electro-optic devices. If READ is true, then the optical exchange/bypass network routes the incoming packet to the bit-rate down converter and switches an empty packet on-line. If the EMPTY and QUEUE bits are both on, the empty packet is replaced with the new packet. To avoid cross-talk and electric field interference at the input of the 2×2 switches, only a single input to each routing switch is used. The new packet is synchronized to the data on the ring by having the local laser operating periodically in phase with the ring data at $\frac{1}{16}$ the repetition rate. In other words, instead of turning on the new packet laser only when the node wants to send out a packet, which would then require a bit-rate switch to align the bits, we

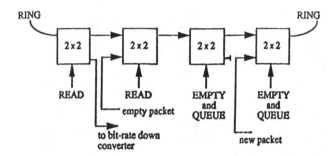

Fig. 5.21 Schematic of the exchange/bypass network, which consists of 2×2 electro-optic switches that reconfigure at the packet rate.

continuously run the new packet laser and block the output when nothing is queued. Therefore, in this case we solve the synchronization problem by throwing away laser power rather than using bit-rate triggering switches.

Once the optical packet reaches its destination, fast gates must demultiplex the packet information to bit rates accessible by electronic shift registers. As discussed in Section 5.1, bit-rate down conversion only requires devices that can operate at the repetition rate of the bits. Although ultrafast soliton-trapping or -dragging logic gates or the all-optical routing switches of Section 5.1 can be used for this function, it may be simpler and cheaper to use 2×2 narrow band devices such as lithium-niobate waveguide switches. Further details on the various modules as well as failure safeguards and the scavenger network node are described in Ref. [5.20].

Ultrafast gates in code-matching logic

The heart of the design is the code-matching logic module that is shown in block diagram form in Fig. 5.22. The first gate of the code-matching logic module removes the 8-bit address header from the packet and performs logic-level and timing restoration. The resulting signal is duplicated and forwarded to two separate selection circuits. The upper AND-gate detects if the packet is empty: an empty packet is assigned an all-one header to maintain clock synchronization. The lower circuit, which consists of an XOR-gate and another AND-gate, detects if the packet should be read. We select the logic functionality for both circuits so that the desired output signal occurs only when all eight output bits are zero, making it simpler to set the threshold. The strobe and XOR-gates are soliton-dragging

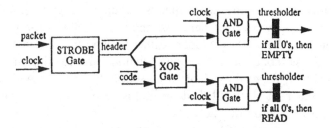

Fig. 5.22 Block diagram of the code-matching logic, which contains
the ultrafast soliton gates. The first two levels of logic use soliton-
dragging logic gates, since they are cascadable and have fan-out,
although their output is in time-shift keyed format. The final stage
consists of soliton-trapping AND-gates, which convert the time-shift
keyed signals to outputs with energy contrast.

logic modules that have fan-out and cascadability (as described in
Section 3.1), since more than one level of logic is used. As the strobe
gate must also provide timing restoration, a prechirper or soliton-
dragging logic gate with an EDFA should be employed, as discussed
in Chapter 4. Also, if a more complicated protocol is desired, we can
add further intermediate stages of soliton-dragging logic elements.
The output of these first stages of logic is in time-shift-keyed format,
but the final stage of AND-gates are soliton-trapping devices (Section
3.4). The soliton-trapping AND-gates are sensitive to the timing of
the input pulse and provide energy contrast at the output, so they can
be used to convert time-shift keying to amplitude-shift keying that is
required for the detectors (cf. Section 3.5).

It is interesting to compare this time-division-multiplexed system
with the dense wavelength-division-multiplexed (DWDM) counterpart
[5.23]. In the simplest DWDM scenario, each node is assigned a frequency and its receiver is tuned to that frequency. An agile, tunable
transmitter at each source is used to communicate with other nodes.
Although the network is a shared multi-access resource, each frequency channel is owned by a single user at each destination. Furthermore, the data destined for different nodes are in different physical
representations.

In our time-multiplexed system, the data for every destination
shares the same physical representation, which makes tasks such as
broadcasting simpler. Sharing the same physical representation also
means that the hardware at each access node can be duplicated.
Instead of a distinct frequency, each node is now assigned a unique
address. The local address code acts as the tuned receiver, and the

agile, tunable transmitter corresponds to different codings of the header in the new packets. Thus, we find that ultrafast bit-rate gates permit us to make a "logical tuner" rather than a physical tuner.

The access node in a 100Gbit/s soliton, slotted ring network exploits the high speeds available from soliton logic gates. A detailed study by Sauer, et al. [5.20], finds several issues that arise in trying to implement the soliton ring network:

(1) in the 100 Gb/s regime, the logic gates should have switching energies at or below a picojoule to maintain the optical power requirements at reasonable levels;

(2) even for a system with just a few ultrafast gates, the logic gates must be cascadable, restoring both timing and amplitude, for a robust implementation;

(3) optical amplifiers with high saturation power ($\sim 0.3\,\text{W}$) are needed for ultrafast, all-optical transport systems;

(4) clock extraction and packet-injection synchronization at the bit-level are major issues for an ultrafast network.

The first two issues have been discussed in Chapter 3, and the last issue is touched on in Chapter 4. There is considerable work devoted to the use of EDFAs for long-distance communication, and the third issue may be solved in the long-haul arena. Thus, the required devices and sub-systems can be crudely demonstrated now, but more development is needed to reach the level of a realistic demonstration.

5.4 Summary

The general issues addressed in considering a few applications of ultrafast gates to telecommunications and optical processing are: what significant tasks can be performed with a handful of ultrafast gates, and what new capabilities do the ultrafast gates bring to the system? It is important to study systems because they provide the specifications and properties that a useful device must exhibit. Also, by looking at the complete system we find where various complexities are traded off. For example, gates with logic and timing restoration often transfer the complexity to the laser or power-supply source, which must provide standard pulses in synchronism with a master clock.

Various implementations of high-speed demultiplexers may be used at the receiving end of a transmission fiber (see Section 5.1), and such devices may bit-rate down-convert a data stream from optical to electronic speeds, or they may provide serial-to-parallel conversion.

Whereas perfectly periodic demultiplexing requires a device with a narrow bandwidth centered around the operating frequency, the demultiplexing of an arbitrary bit pattern requires bit-rate switches. The key challenges for all-optical demultiplexers are operation beyond electronic speeds and operation with switching energies approaching a picojoule to demultiplex arbitrary bit patterns. The application of routing devices, as described in Chapter 2, such as Kerr modulators and nonlinear optical loop mirrors, may solve one or more of these challenges.

The examples in Section 5.2 relate more to processing or computation. The GEO-module is an idealized logical representation of ultrafast, all-optical soliton logic gates that can be used for designing device-independent architectures. In addition to being Boolean and connectivity complete, GEO-modules maximally exploit the device physics and, therefore, lead to minimal component implementation of all-optical functionalities. The counter is the simplest example of a finite-state machine, which, from a computational standpoint, is the next major step toward a computer. Time-domain switches are used in switching systems to interchange time-slots, and simple time-slot interchangers may be used to exchange adjacent packets. Devices that satisfy the GEO-module functionality can, thus, do the first level of ultrafast processing in high-performance systems.

Ultrafast logic gates, such as those described in Chapter 3, have opened the possibility of implementing a local or metropolitan area network with 100 Gbit/s peak data rates and several hundred users. The design in Section 5.3 is a pipelined, feed-forward architecture that exploits the bandwidth while downplaying the latency of fiber logic gates. The ultrafast gates are used sparingly only to select and decode the header, and they permit us to implement a fully digital "logic tuner." The design example discloses several remaining hardware challenges: (1) lower switching energy for logic gates; (2) cascadable logic gates that restore both timing and amplitude; (3) high-saturation power amplifiers; and (4) clock extraction and packet injection synchronized at the bit level.

6

Summary and future prospects

This book focuses on ultrafast, all-optical switching in fibers and attempts to cover the spectrum from basic physical principles to applications. Fibers are an attractive medium for switching because they exhibit interesting physical phenomena such as dispersion, nonlinearity, and Raman gain. Also, optical fiber is one of the very few examples of a medium that behaves one dimensionally since the transverse mode pattern is well confined. Nonlinear guided-wave switches are based on the Kerr effect, and, if these switches are operated in the anomalous group-velocity-dispersion regime, then soliton pulses can also be supported (solitons are preferred for picosecond and femtosecond pulses). The Kerr effect augmented by a phase-matching condition leads to parametric four-wave-mixing, and if anomalous group-velocity dispersion is added, then modulational instability results. Pulses may walk through one another because of group-velocity dispersion (different frequency pulses) or because of birefringence (orthogonally polarized pulses). The Kerr effect between pulses, which is referred to as cross-phase modulation, along with complete walk-through of the pulses leads to integrated phase shifts for pulses and to elastic collisions for solitons. On the other hand, for temporally coincident solitons birefringence and cross-phase modulation lead to soliton-trapping and soliton-dragging.

A variety of routing and logic devices are based on the third-order susceptibility $\chi^{(3)}$ in fibers (see Chapters 2 and 3). Routing switches typically have the control pulse at a different frequency and have multiple output ports, and the decision is represented by the location or position of the data. Operated as a single-input and -output device, a routing gate may function as a saturable absorber or an optical limiter. Examples of routing devices include Kerr modulators, four-wave-mixing gates, and Mach–Zehnder interferometers. On the other hand, digital logic gates represent their decision by a "0" or "1" level and should be regenerative – i.e., the signal pulses are replaced with new power supply pulses that are corrected in amplitude, pulse shape and timing. Soliton logic gates include soliton-dragging, soliton-trapping, and billiard-ball soliton-interaction gates. Soliton-interaction gates illustrate the principles of conservative logic, and soliton-trapping gates are sensitive to the timing at the input and provide an energy contrast at the output. Soliton-dragging logic gates are three-terminal devices with logic-level restoration that satisfy all requirements for digital logic gates (cascadability, fan-out and Boolean completeness). The fiber-nonlinear optical-loop mirror can be configured either as a routing or logical switch, depending on the control and signal orientations.

Because of the long latency typical of all-optical fiber devices, the system architectures should be pipelined and feed-forward. Ultrafast, all-optical switches will be used for serial processing applications that require simple functionalities at terahertz speeds. Demultiplexing at the receiving end of a telecommunications fiber may be one of the first niches for ultrafast devices. In addition, as the bit-rates on long-haul transmission systems continue to increase, the bottleneck for high speed systems will shift to local and metropolitan area networks. A complete paper design, as described for the soliton ring network in Section 5.3, can act as a starting point in recognizing the opportunities and complexities that ultrafast logic gates create. From a computational standpoint, the next step after logic gates is combinatorial and sequential circuits, followed by a finite-state machine that incorporates logic and memory. To illustrate the progression, soliton-dragging logic gates might be used in bit-serial counters and time-slot interchangers. Furthermore, it is also important to introduce logical abstractions such as generalized exclusive-OR (GEO) modules for designing device-independent architectures.

There are several key challenges to building all-optical devices (see Chapter 1), and solutions to these difficulties were discussed throughout the book. A three-terminal device can consist of two inputs at different frequencies or polarizations, which interact through cross-phase modulation, and an output at a separate spatial location. For a cascadable switch (output pulse shape similar to the input) we can use square pulses or complete slip-through between pulses if the device length is shorter than the soliton period Z_0 and the pulse shapes do not distort seriously. Cascadability can also be satisfied by using solitons in the anomalous group-velocity-dispersion regime. Optical devices that exhibit fan-out directly from the physical mechanism generally take advantage of the time axis (e.g., time-shift keying). Otherwise, net gain can be demonstrated by combining an ultrafast device with an amplifier, such as an EDFA. In general, if amplifiers are used then optical limiters may be required to clip the output beyond some level, and saturable absorbers or frequency filters may be needed to block spontaneous emission. Another issue for all-optical devices is reducing the switching energy, and the simplest approach is to use longer interaction lengths and materials with larger nonlinearities. A better approach is to change the switch architecture so that solitons can be used with a "lever arm," as illustrated by the time-domain chirp switch. Finally, temporal jitter and synchronization are always increasingly difficult problems as the speed increases. At the device level these problems can be reduced by correcting the timing of the incoming bit-stream and increasing the timing window accepted by the ultrafast gates. In system applications optical phase-locked loops must be used to synchronize various lasers, and care must be taken to avoid clock skew.

Technological challenges for the future

All-optical devices and systems are still at an infancy stage in their development, and there are several technological areas that require major breakthroughs before the field can thrive. The crucial device issue remains lowering the switching energy so that each device requires less than a picojoule net energy per bit. Novel nonlinear materials must also be studied to make compact devices with reduced latency and increased thermal stability. Perhaps the main missing component for all-optical systems is a compact laser source with average powers approaching a watt that can act as the power supply. Also absent is an all-optical random-access memory, and solutions to this problem have not even been proposed. However, the need for random-access memory in a pipelined machine may be debatable, since the latency in accessing the memory could be detremental. Although commonly overlooked as an "engineering detail," accurate time-synchronization circuits are needed for bit periods approaching a picosecond. Finally, in a broader context, architectures must be crafted that use the switch bandwidth to enhance the capabilities of the system, thereby enabling this ultrafast technology to make an impact.

It is encouraging that progress continues in many of these areas. For example, several groups are studying the nonlinear optical properties of semiconductors and organics below half of the band edge [6.1–6.5], which is the wavelength range where two-photon absorption does not prohibit all-optical switching. The third-order nonlinear optical properties of semiconductors are of interest for making compact, integrable devices where many devices can be grown simultaneously on the same wafer. In addition to exhibiting Kerr nonlinearities, semiconductors are interesting because of unique electric field-dependent behavior [6.6, 6.7], such as the quantum confined Stark effect [6.8]. Semiconductor materials have already been used below half bandgap in devices, as we saw in the hybrid time-domain chirp switch that was discussed in Sections 3.2 and 4.2. Also, an all-optical nonlinear directional coupler using two closely spaced waveguides in AlGaAs material has been demonstrated using 10 ps pulses [6.9]. The GaAs/AlGaAs material system is particularly attractive for all-optical switching for several reasons: (1) by varying the alloy composition the half-bandgap energy can cover the 1.3 to 1.6 μm infrared window that is important for optical communications; (2) the material is almost perfectly lattice matched; and (3) a mature fabrication technology already exists for these alloys.

Although larger nonlinear coefficients are predicted for organics [6.10, 6.11], the material quality, stability and processing techniques are far behind those of semiconductors. For example, the waveguides made in organics still have significant scattering losses, and the longevity of organics under thermal cycling and optical excitation has not been fully explored. Nonetheless, organics may demonstrate unique nonlinear properties because it has been predicted that isolated absorption lines (like atomic resonances) might exist instead of absorption bands (like semiconductors). This means that the operating wavelengths might approach the resonances without deleterious effects from two-photon absorption.

Furthermore, "gap solitons" have been theoretically studied in nonlinear waveguides that consist of layered materials or corrugated surfaces [6.12–6.14]. The "gap soliton" results from the balance of the material nonlinearity and dispersion for frequencies near the stop band (corresponding to the Bragg reflection condition) of the periodic structure. As the index in the waveguide changes with intensity, the gap shifts to different wavelengths: consequently, some wavelengths that were originally reflected shift out of the gap and begin to transmit. Such a slow-wave structure might yield particle-like soliton properties in semiconductors or organics, which could then serve as the second part of the time-domain chirp switch architecture.

Furthermore, if gap solitons are implemented in a semiconductor then the stop band and dispersion could be tuned by an applied electric field.

Work on short-pulse lasers has grown steadily in the last two decades. Although much of the work described in this book used color-center lasers, these lasers are large, cryogenic systems that are pumped by Nd:YAG lasers. Potential candidates for the required compact laser sources include mode-locked semiconductor lasers [6.15], semiconductor pumped solid state lasers [6.16, 6.17], and EDFA lasers [6.18–6.21]. Mode-locked semiconductor lasers still fall short of the desired one-watt average power level, and room temperature, solid state lasers (e.g., Ti:Sapphire and Nd:YAG) have not as yet operated in the telecommunications window around $1.55\,\mu\text{m}$. Recently EDFAs have been incorporated in fiber lasers to generate subpicosecond pulses. Although the power levels also fall short of the watt level, improvements in EDFAs are developing at a rapid pace, and the passively mode-locked EDFA laser work is promising. Furthermore, EDFAs are already making a tremendous impact in long-haul transmission systems, and they may also play a major role in ultrafast switching. One certain use is to loosen loss margins by compensating for coupling, splitting and insertion losses within the system.

The development of novel architectures will need increased interactions between computer designers, systems engineers and device physicists. Although much attention has been directed to simple demultiplexers, interest is beginning to turn toward local and metropolitan area networks. Computer networks, where ultrafast gates may be used for clock synchronization and communication between memory and processors, remain virgin territory. Unlike telecommunication applications that have not as yet exceeded the current device capabilities, current high-performance computer systems are already limited by communication rates between memory and processors. Can a long-latency, ultra-high-speed, logic gate help break the communications bottleneck in computers?

So, where is the field of ultrafast switching devices and systems heading? The field has reached its current status because of a combination of knowledge of photonic switching, short-pulse-generation techniques, nonlinear optics and the optical-fiber technology. Fibers are an almost ideal nonlinear material with long interaction lengths, and fibers have well-understood governing equations as well as a mature fabrication technology. For these reasons fibers will continue to be the testing ground for new all-optical switch architectures.

Furthermore, devices or systems that are longer than a soliton period Z_0 will use solitons for switching and transmission. However, work is likely now to shift toward implementing the fiber gates in more compact form with less switching energy. The ultrafast systems will probably gravitate toward the gain band of EDFAs, perhaps even being driven by EDFA lasers. Potential applications that apply ultrafast gates to local and metropolitan area networks, computer systems, and all-optical random-access memory need to be explored. Although many challenges remain, there are many clever people out there working on the key technologies. Therefore, perhaps the best way to end is to say ... stay tuned.

Basic soliton physics

Solitons in optical fibers can be defined as nonlinear pulses that pro-pagate nearly distortion-free for long distances and that undergo elas-tic collisions. In a linear system, the output is proportional to the input and the response for a sum of inputs is the sum of the indivi-dual responses. On the other hand, when overdriven the nonlinearity of a system becomes apparent and the result increases not only as the sum but also as the product of responses. For example, when an audio speaker is overdriven, then the cone motion is not in direct proportion to the signal and the sound is distorted. Solitons are stable, robust solutions owing to a restoring force that comes from the nonlinearity balancing another mechanism, such as diffraction or dispersion. Although the medium is nonlinear, solitons that undergo elastic collisions pass through each other without exchanging or scattering energy.

Solitons result from many different physical phenomena. Examples include: water wave solitons, ion-plasma-wave solitons, magnetohy-drodynamic solitons, bimolecular polaron solitons, high-intensity shock solitons, nerve-conduction solitons and nonlinear optical soli-tons [A.1]. In 1838 Russell first observed solitary waves in a narrow barge canal and wrote of "a large solitary elevation, a rounded, smooth and well-defined heap of water which continued its course along the channel, apparently without change of form or diminution of speed [A.2]." Russell followed this solitary wave on horseback for one or two miles, eventually losing it in a winding channel. He derived an expression for the velocity of solitary waves in a channel of uniform depth and proposed that the wave does not damp out under idealized conditions. Korteweg and deVries derived in 1895 the hydrodynamic equations of waves that move with velocities propor-tional to their amplitudes and found agreement with the observations [A.3].

In 1965 Zabusky and Kruskal [A.4] discovered that when two or more Korteweg–deVries solitary waves collide they do not break-up and disperse, and they first used the term "soliton" to refer to these particle-like solitary waves. Then, in 1967 Gardner, et al. [A.5], discovered that the Kortweg–deVries equation is analytically solvable. In their 1971 seminal paper, Zhakharov and Shabat [A.6] solved the nonlinear Schrödinger equation (NLSE) using the inverse scattering transform and showed that the solutions were solitons. They also say that the NLSE applies to "a quasi-monochromatic one-dimensional wave in a medium with dispersion and inertialess non-linearity." Hasegawa and Tappert [A.7] in 1973 pointed out that the Zhakharov and Shabat work applied to optical fibers, showed theoretically that an optical pulse in the fiber forms an envelope soliton, and discussed the importance this might have for long-distance communications. Then, in 1980 Mollenauer, et al. [A.8], first experimentally demonstrated soliton propagation in fibers. In the dozen years since, the experimental and theoretical exploration of optical solitons in fibers has reached the state where solitons are now seriously being considered for long-distance telecommunications applications [A.9]. Furthermore, as discussed in this book, in the last few years solitons have been used extensively in ultrafast switching experiments.

Optical solitons are formed when intense pulses propagate in the anomalous (or negative) group-velocity dispersion regime of fibers. Dispersion means that different wavelengths of light travel at different speeds, and anomalous dispersion means that longer wavelengths travel at a slower speed. For example, the dispersion and loss per unit-length of state-of-the-art, non-dispersion shifted fibers is shown in Fig. A.1, where anomalous dispersion corresponds to wavelengths longer than $1.3\,\mu m$. The fiber loss increases at shorter wavelengths because of Rayleigh scattering, and the loss increases at longer wavelengths because of infrared absorption [A.10]. Fortunately, the soliton regime coincides with the minimum loss in fibers around $1.5\,\mu m$, which is one reason that solitons are being considered for long-distance telecommunications.

Since different frequencies propagate at different speeds, dispersion alone tends to broaden any pulse. However, a high-intensity light pulse increases the index-of-refraction and creates a local time-varying index, which corresponds to self-phase modulation. The non-linear index is given by

$$n = n_0 + n_2 I \tag{A.1}$$

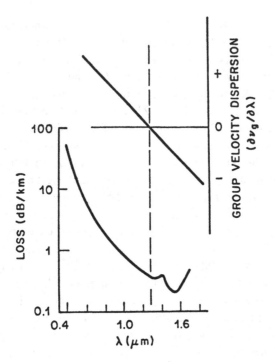

Fig. A.1 Loss and group-velocity dispersion as a function of wave-length for state-of-the-art, non-dispersion shifted fibers.

where the Kerr coefficient n_2 has a value of $n_2 = 3.2 \times 10^{-16} \, \text{cm}^2/\text{W}$ in fused silica. Self-phase modulation has an associated phase change

$$\Delta\phi = \frac{2\pi}{\lambda} L n_2 I, \qquad (A.2)$$

where L is the length of the fiber, that leads to a frequency sweep or chirp across the pulse given by

$$\delta\omega = -\frac{\partial\Delta\phi}{\partial t} = -\frac{2\pi}{\lambda} L n_2 \frac{\partial I}{\partial t} . \qquad (A.3)$$

As shown in Fig. A.2, for a positive value of n_2 self-phase modulation red shifts (shifts to lower frequency) the leading edge and blue shifts (shifts to higher frequency) the trailing edge of the pulse. Therefore, self-phase modulation advances the longer wavelengths while anomalous group-velocity dispersion retards the longer wavelengths, and for the proper intensity and profile of light the two effects cancel each other. The intensity where the two effects exactly balance corresponds to the fundamental $N = 1$ soliton, and at this intensity the pulse propagates without changing shape.

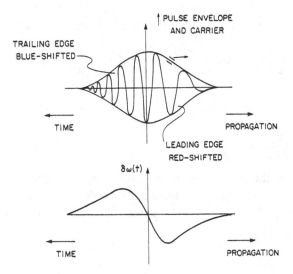

Fig. A.2 Frequency chirping because of self-phase modulation.

Nonlinear Schrödinger equation

The balance between self-phase modulation and group-velocity dispersion can be quantitatively described by the nonlinear Schrödinger equation (NLSE). Consider a single-mode optical fiber in which z corresponds to distance along the fiber. The electric field can be written as

$$\vec{E} \simeq \vec{E}_m(x,y)e^{i(k_0 z - \omega_0 t)}u\left(z, t - \frac{z}{v_{g0}}\right), \tag{A.4}$$

where the first term is the transverse spatial mode pattern, the second term is the carrier frequency and the final term is the envelope function. The group velocity at the carrier frequency is v_{g0}, and $t - z/v_{g0}$ is the retarded time (frame of reference traveling with the pulse). We normalize the envelope function so that $|u|^2$ represents power in the fiber.

In the retarded time coordinates, $u(z,t)$ approximately satisfies the NLSE [A.11, A.12]

$$-i\frac{\partial u}{\partial z} = \frac{\lambda^2 D}{4\pi c}\frac{\partial^2 u}{\partial t^2} + \frac{2\pi n_2}{\lambda A_{\text{eff}}}|u|^2 u, \tag{A.5}$$

where the first term on the right-hand side corresponds to the group-velocity dispersion and the second term corresponds to the nonlinearity. The dispersion parameter D is the change in pulse delay τ_d with change in wavelength λ per unit fiber length and has standard units of ps/(nm km). For a length d of fiber the change in pulse delay is

$$\tau_d = \frac{\partial}{\partial\omega}\left(\frac{1}{v_{g0}}\right)\Delta\omega\, d = \frac{\partial^2\beta}{\partial\omega^2}\Delta\left(\frac{2\pi c}{\lambda}\right)d = -\frac{2\pi c}{\lambda^2}\frac{\partial^2\beta}{\partial\omega^2}\Delta\lambda\, d,$$

(A.6)

and we find

$$D = \frac{\tau_d}{\Delta\lambda\, d} = -\frac{2\pi c}{\lambda^2}\frac{\partial^2\beta}{\partial\omega^2}.$$

(A.7)

We account for the transverse spatial profile by averaging the non-linear effect over the fiber cross-section. If we define the intensity as $I = P/A_{\text{eff}}$ where $P = \int I\, dA$ is the power, then we find the effective area A_{eff} is given by

$$A_{\text{eff}} = \frac{P^2}{\int I^2\, dA}.$$

(A.8)

The NLSE can be simplified by using standard soliton units [A.11]. We normalize the distances, times and powers to z_c, t_c and P_c, where

$$\frac{t_c^2}{z_c} = \frac{\lambda^2 D}{2\pi c}$$

(A.9)

and

$$P_c z_c = \frac{\lambda A_{\text{eff}}}{2\pi n_2}.$$

(A.10)

The length z_c is the distance at which a low power pulse confined to t_c pulse width begins to spread by dispersion, and this is related to the so-called soliton period Z_0 by $Z_0 = \frac{1}{2}\pi z_c$. The power P_c (which equals the $N = 1$ soliton power P_1) gives one radian of nonlinear phase shift at a distance z_c and is the peak power at which the non-linearity and dispersion balance. The resulting normalized NLSE in retarded time coordinates is

$$-i\frac{\partial u}{\partial z} = \frac{1}{2}\frac{\partial^2 u}{\partial t^2} + |u|^2 u.$$

(A.11)

Notice that the above normalizations have an arbitrariness in that there are only two relations to be satisfied by three quantities. Therefore, one of the quantities can be chosen arbitrarily; e.g., in the experiments we typically select the time scale to correspond to the laser pulse width and then the other variables are then constrained. More generally, this arbitrariness means that there are scaling rules for any solution of Eq. (A.11). For example, if $u(z, t)$ is a solution to Eq. (A.11), then $Ru(R^2z, Rt)$ is also a solution where R can be an arbitrary real number. There is also a family of solutions that are frequency-shifted versions of $u(z, t)$, which, with a frequency shift of $-\Omega$, are given by $u(z, t - \Omega z) \exp\{i(\Omega t - \frac{1}{2}\Omega^2 z)\}$.

The NLSE of relation (A.11) is integrable because it can be solved by the inverse scattering technique, and, therefore, has an infinite number of conserved quantities [A.6]. The three lowest conserved quantities are [A.13]

$$C_1 = \int_{-\infty}^{\infty} |u(z, t)|^2 \, dt \tag{A.12}$$

$$C_2 = i \int_{-\infty}^{\infty} \left(u^* \frac{\partial u}{\partial t} - u \frac{\partial u^*}{\partial t}\right) dt \tag{A.13}$$

and

$$C_3 = \int_{-\infty}^{\infty} \left(\left|\frac{\partial u}{\partial t}\right|^2 - |u|^4\right) dt. \tag{A.14}$$

The physical interpretation of these conserved quantities depends on the particle or wave picture of solitons. In the wave picture, relation (A.12) corresponds to conservation of energy and (A.13) corresponds to conservation of the mean frequency as weighted by the intensity. On the other hand, in a particle picture of the soliton C_1 corresponds to conservation of mass, C_2 corresponds to conservation of momentum and C_3 corresponds to the Hamiltonian or conservation of energy.

Soliton solutions

The canonical single-soliton solution to Eq. (A.11), which corresponds to an $N = 1$ fundamental soliton, has the form

$$u(z, t) = \text{sech}\, t \, e^{\frac{1}{2}iz}. \tag{A.15}$$

The normalized energy is given by

$$E_s = \int |u|^2 \, dt = 2,$$

and the full-width-at-half-maximum pulse width in the normalized units is $\tau/t_c = 2\cosh^{-1}\sqrt{2} = 1.763$. An important point about relation (A.15) is that the phase of the pulse depends on the position along the fiber and not on the temporal position of the pulse. Therefore, the fundamental soliton has a uniform phase shift and can switch as a unit without pulse break-up. The factor of half in the exponent of (A.15) can be interpreted as the soliton averaging the phase factor over the entire pulse. The Fourier transform, which can be defined as

$$u(t) = \int d\Omega \, \tilde{u}(\Omega) \exp(-i\Omega t), \qquad (A.16)$$

gives the soliton-frequency spectrum corresponding to Eq. (A.15) as

$$\tilde{u} = \tfrac{1}{2} \operatorname{sech} \tfrac{1}{2}\pi\Omega, \qquad (A.17)$$

where $\Omega = (\omega - \omega_0) t_c$ is the frequency separation.

The fundamental soliton can be written in more general form than (A.15) by using the scaling rules to yield [A.14]

$$u(z,t) = A \operatorname{sech}(At - q) \exp\{-i(\Omega t + \phi)\}. \qquad (A.18)$$

The amplitude of the soliton is A, the energy is $2A$, the mean frequency is Ω, the phase is ϕ and the mean time is q/A. A positive value of Ω corresponds to a positive frequency displacement. The displacement and the phase evolve along the fiber according to

$$\frac{dq}{dz} = -A\Omega; \qquad \frac{d\phi}{dz} = \tfrac{1}{2}(A^2 - \Omega^2). \qquad (A.19)$$

Therefore, at some location z in the fiber, a soliton is completely specified by the values of its four parameters A, Ω, q and ϕ.

In addition to the fundamental soliton $(N = 1)$, there are also a continuum of multiple-soliton solutions that obey the NLSE [A.12]. Unlike the fundamental soliton that behaves as a unit and represents a balance between dispersion and nonlinearity, the higher-intensity and -order solitons change shape as they propagate down the fiber since the two counteracting forces overshoot and undershoot. For example, a higher- or multiple-order soliton tends to compress at first because the self-phase modulation outweighs the group-velocity dispersion. However, as the pulse narrows, the bandwidth of the pulse increases and the dispersive effects become stronger. The general N-soliton solution is characterized by $4N$ parameters: A_j, Ω_j, q_j and ϕ_j $(j = 1, \ldots, N)$ [A.15]. Of particular interest are bound soliton solutions where all the solitons share a common frequency $\Omega_j = \Omega$ and, consequently, a common velocity. The bound multi-solitons evolve periodically and the patterns corresponds to constructive and destructive interference between the pulses. As an example, the general two-soliton function with arbitrary Ω_1 and Ω_2 is described by Eq. (C.3) in Appendix C. If the position and velocities of the two solitons are equal initially $(\Omega_1 = \Omega_2; \; q_1 = q_2)$, then $N = 2$ solitons with

$(\eta_1 + \eta_2 = 2)$

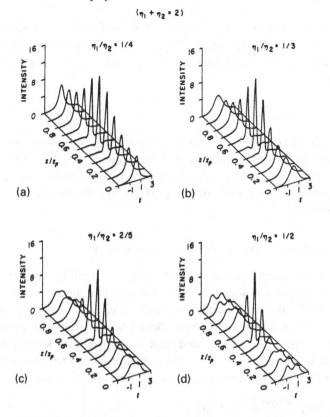

Fig. A.3 Symmetric two-soliton functions versus normalized time t and normalized distance z/z_p, where z_p is the repetition period. The two bound solitons have equal input frequencies and positions. The energy for the four cases is held fixed ($\eta_1 + \eta_2 = 2$), while the ratio η_1/η_2 is varied. (a) $\eta_1/\eta_2 = 0.25$; (b) $\eta_1/\eta_2 = 0.33$; (c) $\eta_1/\eta_2 = 0.4$; and (d) $\eta_1/\eta_2 = 0.5$.

eigenvalues η_1 and η_2 ($A_j = 2\eta_j$) can be written as [A.12]

$$u = \left\{ \frac{4\eta_1(\eta_1+\eta_2)}{|\eta_2-\eta_1|} \exp\{2i\eta_1^2 z\} \right.$$

$$\left. \times \left(\cosh 2\eta_2 t + \frac{\eta_2}{\eta_1} \cosh 2\eta_1 t \exp\{2i(\eta_2^2 - \eta_1^2)z\} \right) \right\}$$

$$\Bigg/ \left\{ \cosh\{2(\eta_1+\eta_2)t\} + \left(\frac{\eta_1+\eta_2}{\eta_2-\eta_1}\right)^2 \cosh\{2(\eta_2-\eta_1)t\} \right.$$

$$\left. + \frac{4\eta_1\eta_2}{(\eta_2-\eta_1)^2} \cos\{2(\eta_2^2-\eta_1^2)z\} \right\} \quad \text{(A.20)}$$

Figure A.3 shows a plot of relation (A.20) for different ratios of η_1/η_2, and we find different shapes and evolutions for different parameters. These bound two-soliton functions repeat with a period given by

$$z_\mathrm{p} = \frac{\pi}{(\eta_2+\eta_1)(\eta_2-\eta_1)}$$

Also, relation (A.20) shows that because of the interference between the bound fundamental solitons that the phase across the pulse now depends explicitly on the temporal position of the pulse. Therefore, higher-order solitons turn out not to be interesting for switching applications.

Robustness and asymptotic soliton solution

An intuitive way of understanding the stability of fundamental solitons is to consider the dispersion relation shown in Fig. A.4, where we plot the wave-number versus frequency for the unperturbed soliton and for a linear dispersive field [A.14]. From the canonical solution in Eq. (A.15), we find that a fundamental soliton has a wavevector given by $k_\mathrm{s} = \frac{1}{2}$, which is independent of frequency. On the other hand, a low-power dispersive wave is of the form

$$u_\mathrm{disp} \propto \exp\{\mathrm{i}(kz - \omega t)\}, \tag{A.21}$$

and, if we neglect the nonlinear term in Eq. (A.11), then we obtain a parabolic-dispersion relation $k_\mathrm{d} = -\frac{1}{2}\omega^2$. In general, a dispersive wave is phase mismatched from the soliton and, therefore, does not disturb the soliton. Solitons are deleteriously affected by perturbations in the fiber or transmission line parameters that make up the wave-vector difference between the two curves in Fig. A.4. For example, a perturbation of the form $\cos k_\mathrm{p} z$ couples power out of the soliton only at angular frequencies given by $\pm\sqrt{2k_\mathrm{p}-1}$, with a rate of loss proportional to the spectral intensity of the soliton at that frequency. Thus, solitons are vulnerable only to perturbations with $k_\mathrm{p} \sim \frac{1}{2}$ (a period of 4π or $8Z_0$) because then the dispersive wave is both phase- and group-velocity matched to the soliton [A.14]. A long period perturbation, whose wave-number has absolute value less than one half, cannot couple energy out of the soliton. As a consequence, solitons are stable against slow parameter variations in the propagation equation. A short period perturbation causes little energy loss because the soliton has little energy at the phase-matched frequency.

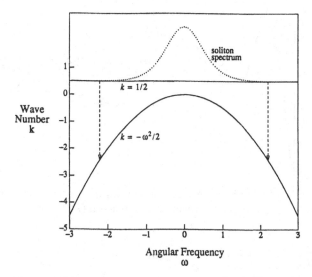

Fig. A.4 Dispersion curves for a soliton ($k_s = \frac{1}{2}$) and for linear waves ($k_d = -\frac{1}{2}\omega^2$) [A.14].

Solitons are eigenfunctions of the nonlinear fiber equations in the sense that an arbitrary input with sufficient energy will evolve into a soliton. Although the solution for a general input has to be determined numerically, Satsuma and Yajima [A.15] show analytic solutions for hyperbolic secant inputs. An input of the form

$$u_i = (1+a)\operatorname{sech} t \qquad (-\tfrac{1}{2} \leq a \leq \tfrac{1}{2}) \tag{A.22}$$

at $z = 0$ will evolve asymptotically into a soliton given by

$$|u_s(z \to \infty)| \simeq (1+2a)\operatorname{sech}\{(1+2a)t\}. \tag{A.23}$$

Note that this soliton has a constant π-area: for $a > 0$ the pulse narrows, and for $a < 0$ the pulse broadens. The tendency of a pulse to approach a standard, constant-area soliton is important for logic-level restoration in logic gates and optical regenerators that use optical amplifiers.

The input pulse evolves into a soliton by stripping off a fraction of energy $[a/(1+a)]^2$ that is lost to a dispersive or radiative field. This radiation can interfere with the soliton, particularly when the wavevector difference is $\sim \frac{1}{2}$. For example, Gordon [A.14] has shown that an input $(1+a)\operatorname{sech} t$ approaches an asymptotic soliton and behaves to first order in a at the center of the pulse $t = 0$ as

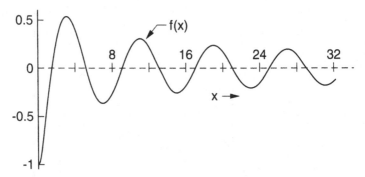

Fig. A.5 Impulse response function for a gain perturbation. The function $f(x)$ can be approximated by relation (A.25).

$$\frac{|u|^2}{|u_s|^2} \cong 1 + f(x)\left[1 - \left(\frac{1+a}{1+2a}\right)^2\right]; \qquad x = (1+2a)^2\frac{z}{Z_0},$$

(A.24)

where the function $f(x)$ is plotted in Fig. A.5. We find that the beating has a periodicity of $\sim 8Z_0$, as we expected from the phase-matching condition of Fig. A.4. Gordon also finds empirically that the impulse response for a perturbation fits the function

$$f(x) \cong (1+x^{1.2})^{-1/2.4}\sin\left(\tfrac{1}{4}\pi\frac{x^2-2}{x+1}\right).$$

(A.25)

Similar oscillatory curves were shown numerically first by Satsuma and Yajima [A.15].

Modulational instability for quasi-cw inputs

Another evidence of the eigenfunction nature of solitons is that quasi-cw inputs also develop into a train of solitons in a process called modulational instability (MI) [A.17–A.19]. MI can be considered as a parametric four-wave-mixing process in which the non-linearity explicitly enters the phase-matching condition. To derive the dispersion relation and gain bandwidth for MI, we assume a cw input plus a perturbation field of the form

$$u(z,t) = e^{ia^2 z}[a + v(z,t)],$$

(A.26)

where $|v| \ll |a|$, a is real and $ae^{ia^2 z}$ is a cw solution of the NLSE in (A.11). Note in relation (A.26) that the phase between the cw pump and the perturbation is important and the phase difference determines whether a perturbation grows or decays. Introducing (A.26) into (A.11) and retaining terms only to lowest order in v we obtain

$$-i\frac{\partial v}{\partial z} = \frac{1}{2}\frac{\partial^2 v}{\partial t^2} + a^2 v + a^2 v^*. \tag{A.27}$$

If we assume that the perturbation is frequency shifted by $\pm\Omega$ from the input

$$v(z,t) = A(z)e^{-i\Omega t} + B^*(z)e^{i\Omega t}, \tag{A.28}$$

then Eq. (A.27) yields two coupled equations

$$\left(\frac{1}{2}\Omega^2 - a^2 - i\frac{\partial}{\partial z}\right)A = a^2 B; \qquad \left(\frac{1}{2}\Omega^2 - a^2 + i\frac{\partial}{\partial z}\right)B = a^2 A. \tag{A.29}$$

Furthermore, if we assume an exponentially behaving amplitude $A \sim e^{\kappa z}$, then combining Eqs. (A.29) we obtain the dispersion relation

$$\kappa = \pm\Omega\sqrt{a^2 - (\tfrac{1}{2}\Omega)^2}. \tag{A.30}$$

Both signs of κ must be retained since both growing and decaying solutions are generally required to match the initial conditions. Therefore, the sideband frequencies $\pm\Omega$ grow in the anomalous group-velocity dispersion regime for frequencies in the range $0 \leqslant \Omega \leqslant 2a$, and the maximum gain occurs at $\Omega_{max} = \sqrt{2}a$.

In Fig. A.6 we plot the MI gain coefficient for different pump amplitudes, and we find that the bandwidth increases with increasing pump amplitude. The widening of the gain window can be understood intuitively in terms of an area argument. For a beat frequency Ω the period is $2\pi/\Omega$, and the area in one period for a cw of amplitude a is $2\pi a/\Omega$. As we found from relation (A.23), the asymptotic soliton will have an area of π in normalized units. If we require that the area in one period of the cw be at least enough to evolve into a soliton, then we obtain

$$\text{area} = \frac{2\pi a}{\Omega} \geqslant \pi, \tag{A.31}$$

i.e.,

$$\Omega \leqslant 2a.$$

Therefore, the maximum detuning frequency is such that the area in one period is equal to that of a fundamental soliton.

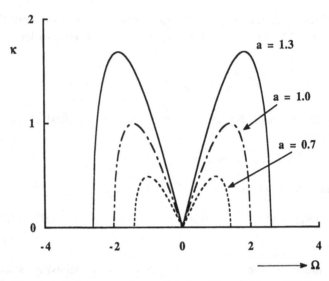

Fig. A.6 Gain spectra for modulational instability at different pump power levels.

We now return to the gain function for the perturbation field v of Eq. (A.27), where we will define the gain $G = |v_{out}|^2/|v_{in}|^2$. The growth rate will depend on the initial condition and whether one or both sidebands are excited. For an incident perturbation of the form $\cos \Omega t$, the gain coefficient as a function of distance in the fiber at the pulse center $t = 0$ is

$$G = 1 + \left(\frac{2a}{\Omega} \sinh \kappa z \right)^2. \qquad (A.32)$$

On the other hand, for a single sideband input of the form $e^{i\Omega t}$ the gain coefficient at that sideband is given by

$$G = 1 + \left(\frac{a^2}{\kappa} \sinh \kappa z \right)^2. \qquad (A.33)$$

In the limit of $\kappa z \gg 1$, the asymptotic form of relation (A.32) is

$$\left(\frac{a}{\Omega} \right)^2 \exp 2\kappa z$$

and the asymptotic form of relation (A.33) is

$$\left(\frac{a^2}{2\kappa} \right)^2 \exp 2\kappa z.$$

Soliton self-frequency shift effect

Thus far we have only treated the simple, unperturbed NLSE of Eq. (A.11), which is an accurate model for short lengths of optical fibers. If wide pulses ($\tau \gtrsim 10\,\text{ps}$) are used, as in long-haul communications, then loss is usually the first term that is supplemented. We can include loss in Eq. (A.11) by adding a term $-i\Gamma u$ to the right-hand side, where $2\Gamma = \alpha_{\text{eff}} z_{\text{c}}$ and α_{eff} is the net (loss minus gain) absorption coefficient [A.11]. However, if short pulses ($\tau \lesssim 1\,\text{ps}$) are used as in ultrafast switching, then the major perturbation to Eq. (A.11) comes from the Raman effect or so-called soliton self-frequency shift (SSFS). Raman effects cause a continuous downshift of the mean frequency of pulses propagating in optical fibers. This is particularly important for short pulses because the effect varies roughly with the inverse fourth power of the pulse width. To understand intuitively the SSFS, consider the Raman gain spectrum [A.20] in fused silica fibers (Fig. A.7). Because of the finite slope down to zero frequency, the soliton can self-induce gain for the lower-frequency part of its spectrum at the expense of the higher-frequency part. Therefore, the mean soliton frequency shifts because of a frequency-dependent gain or loss.

The nonresonant Raman effect in fibers is equivalent to a time-dependent nonlinearity, and there are two components to the Kerr nonlinearity n_2 in relation (A.1). About four-fifths of the n_2 is an electronic, instantaneous nonlinearity caused by ultraviolet resonances, while about one-fifth of n_2 arises from Raman-active vibrations. The imaginary part of this latter contribution corresponds to the Raman gain spectrum in Fig. A.7 and is time dependent with an average delay of ~ 3 to $6\,\text{fs}$ in fused silica (even longer when fiber is doped with germanium). The simplest way to introduce Raman effects into the NLSE of Eq. (A.11) is to modify the nonlinear term to describe a delayed response of the form [A.21]

$$|u|^2 u \rightarrow u(t) \int \mathrm{d}s\, f(s)\, |u(t-s)|^2. \qquad (A.34)$$

The function $f(s)$ is real if there are no losses other than the Raman type, and to ensure causality $f(-|s|) = 0$ while $\int f(s)\, \mathrm{d}s = 1$ to recover Eq. (A.11) for sufficiently short delays (or long pulses). For pulses much longer than the response time (i.e., $\tau > 100\,\text{fs}$) it is generally adequate to Taylor expand Eq. (A.34) as

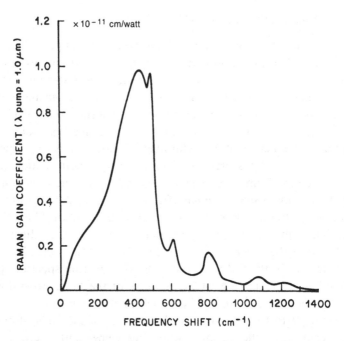

Fig. A.7 Raman gain spectrum in fused silica fibers [A.20].

$$\int \mathrm{d}s \, f(s)|u(t-s)|^2 \rightarrow |u(t)|^2 - t_\mathrm{d}\frac{\partial}{\partial t}|u(t)|^2 + \cdots \qquad (A.35)$$

where t_d is proportional to the slope of the Raman gain curve near zero frequency. For example, in fused silica $t_\mathrm{d} \cong 6[\mathrm{fs}]/t_\mathrm{c}$ [A.22].

Gordon [A.21] has derived analytic formulas to describe the shift of the soliton center frequency as a function of distance in the fiber. He finds for fused silica fibers that

$$\frac{\mathrm{d}\nu_0}{\mathrm{d}z}[\mathrm{THz/km}] = -\frac{10^5\lambda^2 D}{16\pi c t_\mathrm{c}^3}\int_0^\infty \mathrm{d}\Omega \, \frac{\Omega^3 R(\Omega/2\pi t_\mathrm{c})}{\sinh^2 \frac{1}{2}\pi\Omega} , \qquad (A.36)$$

where λ, D, c and t_c are in units of centimeters and picoseconds. The function R is the Raman loss spectrum normalized to the peak value of 0.492 at 13.2 THz, and it can be approximated by a straight line $R(\nu) \simeq 0.492(\nu/13.2)$, where ν is in terahertz. For example, if we use $t_\mathrm{c} = \tau/1.763$, $\lambda = 1.5\,\mu\mathrm{m}$ and $D = 15\,\mathrm{ps/(nm\,km)}$, then with the linear approximation Eq. (A.36) reduces to

$$\frac{\mathrm{d}\nu_0}{\mathrm{d}z}[\mathrm{THz/km}] = \frac{0.0436}{\tau^4} . \qquad (A.37)$$

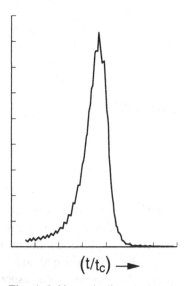

$$\left(t/t_c\right) \longrightarrow$$

Fig. A.8 Numerically calculated pulse shape after propagating in a fiber with SSFS. The input intensity is a squared hyperbolic secant function.

The fourth power of pulse width results since the soliton's peak power scales as τ^{-2}, the spectral width of the soliton is proportional to τ^{-1}, and the Raman gain coefficient increases linearly with increased frequency separation, which also scales as the spectral width. As discussed in more detail in Section 4.1, the SSFS is important for terabit-rate switching and SSFS means that $\tau \gtrsim 500\,\mathrm{fs}$ pulses must be used for fiber lengths of $\sim 300\,\mathrm{m}$.

The SSFS term in Eq. (A.35) is an asymmetric operator that reshapes the pulse as well as shifting its frequency. For example, Fig. A.8 shows a numerical simulation of a short squared hyperbolic secant pulse after propagating in a fiber with SSFS, and we find that the trailing edge of the pulse is sharpened because of SSFS [A.23]. In addition, the asymmetric operator gives a different frequency kick to different pulse sizes. As we discussed earlier, a higher-order or multi-soliton function can correspond to the bound solution for several fundamental solitons. In the presence of SSFS, each of the solitons obtain a different velocity, thereby destroying the bound state [A.24]. Consequently, for short pulse propagation and fiber lengths close to or larger than a soliton period, the fundamental soliton is the only stable solution.

Solitons in birefringent fibers

In the previous appendix, we assumed that the fiber contains only a single mode, which is governed by the nonlinear Schrödinger equation (NLSE). However, in general, even single-mode fibers are bimodal because of birefringence; i.e., the two principal axes in the fiber have indices n_1 and n_2, and the birefringence corresponds to the difference $\Delta n = n_1 - n_2$. Birefringence is unavoidable and can be stress-induced or result from geometric effects. Polarization-maintaining fibers have a birefringence that is a few times 10^{-4}, while moderately birefringent fibers have a birefringence that is a few times 10^{-5}. Standard fibers typically have $\Delta n \sim 10^{-6} - 10^{-5}$, and carefully manufactured telecommunications fibers can have lower birefringences of $\Delta n < 10^{-6}$ by spinning the glass preform during the fiber-pulling process. If the birefringent beat length $L_b = \lambda/\Delta n$ is shorter than the twist length L_t, then a single polarization can be preserved.

Birefringence leads to walk-off between pulses and pulse broadening. In the limit that $L_t < L_b$ and the two fiber axes are scrambled, pulses broaden because of polarization mode dispersion, which reflects the differential delay time between the axes. Poole [B.1] shows for long fibers that the differential delay δt between the two principal axes has a Gaussian probability distribution. This distribution has an expected value $\langle \delta t \rangle = 0$ and a variance $\langle \delta t^2 \rangle = \Delta \beta^2 z/h$, where z is the fiber length, h is the rate at which light is scattered between the axes and $\Delta \beta = d(\beta_1 - \beta_2)/d\omega$ is the difference in group delay per unit length. Mollenauer, et al. [B.2], find that solitons of any pulse width can avoid excess broadening from a randomly varying birefringence. The condition for solitons to prevent broadening is that the fiber's polarization dispersion parameter (in ps/km$^{1/2}$) is less than $\sim 0.3D^{1/2}$, where D is the dispersion parameter in ps/(nm km). However, their numerical simulations also show that polarization dispersion produces a significant amount of dispersive wave radiation from the soliton.

For the switching applications such as soliton dragging and trapping, we use fibers that maintain their axes ($L_b < L_t$) and have a steady birefringence. A uniform velocity difference leads to walk-off between the pulses, which in turn limits the interaction length between orthogonally polarized pulses. However, solitons can compensate for moderate birefringence through mutual nonlinear interaction via cross-phase modulation. In soliton dragging, cross-phase modulation and walk-off lead to an asymmetric interaction between the pulses, which results in a shift of the pulse-center frequencies. A special case of soliton dragging is soliton trapping, when the two polarization components become "self-trapped" in the time domain. In soliton trapping the two polarizations shift their frequencies and, through the group-velocity dispersion, their speeds to make up the velocity difference because of birefringence. Therefore, the Kerr effect, which stabilizes solitons against spreading because of dispersion, also stabilizes solitons against broadening and splitting because of birefringence.

Coupled nonlinear Schrödinger equations

The pulse propagation in a birefringent optical fiber is described by the coupled nonlinear Schrödinger equations (CNLS). Using the standard soliton normalizations that were described in Appendix A, Menyuk [B.3, B.4] derives the CNLS as

$$-i\left(\frac{\partial u}{\partial z} + \delta\frac{\partial u}{\partial t}\right) = \frac{1}{2}\frac{\partial^2 u}{\partial t^2} + |u|^2 u + \tfrac{2}{3}|v|^2 u + \tfrac{1}{3}v^2 u^* \exp(-iR\delta z),$$

$$(B.1)$$

$$-i\left(\frac{\partial v}{\partial z} - \delta\frac{\partial v}{\partial t}\right) = \frac{1}{2}\frac{\partial^2 v}{\partial t^2} + |v|^2 v + \tfrac{2}{3}|u|^2 v + \tfrac{1}{3}u^2 v^* \exp(+iR\delta z),$$

$$(B.2)$$

where u and v are the envelope functions along the two axes. The left-hand side of the equations represents the pulse walk-off because of birefringence, and we select a normalized coordinate system traveling at the mean velocity between the axes. The terms on the right-hand side correspond to group-velocity dispersion, self-phase modulation, cross-phase modulation and coherent four-wave-mixing or phase conjugation. For linearly birefringent fibers that behave isotropically, the cross-phase modulation coefficient is two-thirds of self-phase modulation. The normalized birefringence is given by

$\delta = \pi \Delta n (\tau/1.763)/\lambda^2 |D|$, and the normalized wavevector is $R = 8\pi c t_c/\lambda$.

The final coherence term is responsible for the polarization instability in low-birefringence fibers that was described in Section 2.3, but in soliton-dragging and -trapping experiments this term turns out to be inconsequential. As Menyuk shows with numerical examples [B.3, B.4], for most experimental parameters the coherent term is rapidly oscillating and averages to zero. Therefore, the CNLS reduce to

$$-i\left(\frac{\partial u}{\partial z} + \delta \frac{\partial u}{\partial t}\right) = \frac{1}{2}\frac{\partial^2 u}{\partial t^2} + |u|^2 u + \tfrac{2}{3}|v|^2 u, \tag{B.3}$$

$$-i\left(\frac{\partial v}{\partial z} - \delta \frac{\partial v}{\partial t}\right) = \frac{1}{2}\frac{\partial^2 v}{\partial t^2} + |v|^2 v + \tfrac{2}{3}|u|^2 v. \tag{B.4}$$

Note that for these equations the photon number along each axis is conserved; i.e., $\int |u|^2 \, dt$ and $\int |v|^2 \, dt$ are constants. Also, the interaction through cross-phase modulation is now independent of the phase difference between u and v, which is important for designing phase-independent, all-optical switches.

Zakharov and Shabat [B.5] showed the nonlinear Schrödinger equation of Eq. (A.11) to be integrable by solving it using an inverse scattering transform (cf. Appendix A). Strictly speaking, however, the equations that describe a birefringent optical fiber are not integrable. This nonintegrability occurs because in linearly birefringent fibers the cross-coupling between modes is only two-thirds as strong as self-coupling [B.6]. As illustrated by soliton dragging and trapping, one consequence of the nonintegrability is that inelastic collisions are possible between orthogonally polarized pulses. Furthermore, in switching schemes where pulses pass through one another (cf. Sections 2.4 and 5.1), Menyuk [B.6] predicts that "shadows" may result along the orthogonal axis. For example, when the switching pulse moves through the signal pulse, the switching pulse not only experiences a phase shift but also captures a portion of the orthogonally polarized signal pulse. Therefore, after two orthogonally polarized pulses pass completely through each other, the nonintegrability leads to a change in polarization state. However, as discussed in Section 4.1 and further detailed below, this inelasticity is small compared with the effect of an incomplete walk-through or inhomogeneity during the collision process, as experienced during soliton dragging or trapping. The CNLS are usually studied numerically both because of the nonintegrability and because the equations are much more complicated than the single-axis NLSE.

Numerical solutions for soliton trapping

Menyuk [B.3, B.4] first predicted numerically that soliton trapping should occur for the proper range of birefringences. To determine the amplitude needed to stabilize injected pulses against splitting for a given δ value, Menyuk solved Eqs. (B.3) and (B.4) using a split-step Fourier transform technique. The simplest case corresponds to equal excitation along the two axes, in which case we find from symmetry that

$$t_{max}^v(z) = -t_{max}^u(z), \tag{B.5}$$

where t_{max}^v is the temporal coordinate of the maximum value of $v(z, t)$. If the two pulses do not overlap or the birefringence is too strong, then $t_{max}(z)$ increases without bound. However, if soliton trapping occurs, then $t_{max}(z)$ is bounded. We also define the frequency centroids

$$\omega_{cent}^u(z) = \int_{-\infty}^{\infty} d\omega \, \omega \, |\tilde{u}(\omega, z)|^2 \Big/ \int_{-\infty}^{\infty} d\omega \, |\tilde{u}(\omega, z)|^2, \tag{B.6}$$

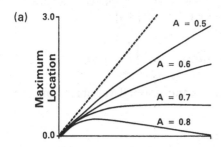

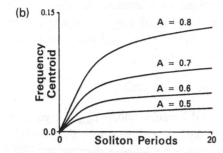

Fig. B.1 Plot of (a) the temporal position of the pulse peak and (b) the frequency centroid as a function of distance along the fiber measured in soliton periods. The net amplitude A is varied and the normalized birefringence is $\delta = 0.15$ [B.4].

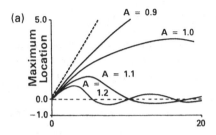

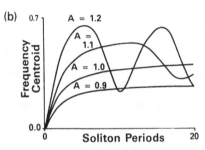

Fig. B.2 Plot of (a) the temporal position of the pulse peak and (b) the frequency centroid as a function of distance along the fiber measured in soliton periods. The net amplitude A is varied and the normalized birefringence is $\delta = 0.5$ [B.4].

$$\omega_{\text{cent}}^{v}(z) = \int_{-\infty}^{\infty} d\omega \, \omega |\tilde{v}(\omega, z)|^2 \Big/ \int_{-\infty}^{\infty} d\omega \, |\tilde{v}(\omega, z)|^2, \qquad (\text{B.7})$$

where $\tilde{u}$ and $\tilde{v}$ are the Fourier transformed variables, and from symmetry we find that

$$\omega_{\text{cent}}^{u}(z) = -\omega_{\text{cent}}^{v}(z). \qquad (\text{B.8})$$

As the value of the birefringence increases, the intensity required for soliton trapping increases. In Fig. B.1 we see the frequency centroid and temporal maximum for $\delta = 0.15$ and different values of A, where the input at $z = 0$ is of the form

$$u(z = 0, t) = v(z = 0, t) = \frac{A}{\sqrt{2}} \operatorname{sech} t. \qquad (\text{B.9})$$

There is a threshold of $A \sim 0.7$ below which t_{max}^{u} increases steadily while ω_{cent}^{u} appears to reach an asymptotic value. Beyond this threshold, the pulses become mutually bound at a distance of about $20Z_0$ and both t_{max}^{u} and ω_{cent}^{u} oscillate over a longer-length scale than shown in Fig. B.1. The behavior becomes sharper and the length scale reduces when we increase the birefringence to $\delta = 0.5$ in Fig.

B.2. Now the threshold value is $A \sim 1.0$ and the oscillations that occur for $A > 1.0$ are clearly visible. Figure B.2 (b) shows that there is a maximum frequency excursion when $A \simeq 1.2$ and $L \simeq 5Z_0$, which are the parameters chosen for the soliton-trapping experiments of Section 3.4. Menyuk also finds that for $\delta \gtrsim 0.75$ the threshold rises to $A \gtrsim 1.5$; however, for short pulses $A \gtrsim 1.5$ higher-order solitons are not stable because of soliton self-frequency shift effects (cf. Appendix A).

Chirp mechanisms in soliton dragging

The curves in Figs. B.1 and B.2 that do not correspond to trapped pulses still show a frequency and temporal change caused by an orthogonally polarized pulse. Therefore, these curves are an example of soliton dragging, which is a more general case than soliton trapping for a wider range of amplitudes A and δ values. Soliton dragging can be understood in terms of a frequency chirp resulting from cross-phase modulation and walk-off because of birefringence, as we discussed in Sections 3.1 and 3.2. Cross-phase modulation during the first one or two walk-off lengths in a birefringent optical fiber causes most of the frequency shift that translates into a time shift after propagating in a dispersive delay line. An intuitive picture of the chirp mechanisms was given in Section 4.2 (Fig. 4.1), where a substantial frequency shift is shown to result in soliton dragging because each pulse experiences only part of the complete collision. Furthermore, introducing gain or loss during the pulse interaction asymmetrizes the pulse walk-off and, consequently, increases the temporal window over which soliton dragging can result in a time shift. The equations that can be used to model soliton dragging and generate the curves in Fig. 4.1 are now derived [B.7].

To calculate the frequency shift of the control pulse because of cross-phase modulation, we return to the CNLS of relations (B.3) and (B.4). If we assume that the control pulse travels along the slow axis, u, then we find from Ref. [B.4] that its frequency centroid is given by

$$\omega_c = i \int_{-\infty}^{\infty} dt \left(u^* \frac{\partial u}{\partial t} \right) \Big/ \int_{-\infty}^{\infty} dt \, |u|^2. \tag{B.10}$$

The derivative of Eq. (B.10) with respect to z can be calculated by using Eqs. (B.3) and (B.4)

$$\frac{d\omega_c}{dz} = B \int_{-\infty}^{\infty} dt \left(|v|^2 \frac{\partial}{\partial t} |u|^2 \right) \bigg/ \int_{-\infty}^{\infty} dt \, |u|^2, \tag{B.11}$$

where the cross-phase modulation coefficient for linearly birefringent fibers is $B = \frac{2}{3}$.

We assume that the two injected pulses are proportional to the shape of a fundamental soliton ($u = A_c \text{sech}(t - \delta z)$, $v = A_s \text{sech}(t + \delta z)$) and that the two pulses interact before their shapes change substantially. Then, Eq. (B.11) can be integrated over t and then over z to obtain $\Delta\omega_c$.

$$\Delta\omega_c = F(z) - F(z_0); \quad F(z) = \frac{BA_s^2}{\delta} \frac{2\delta z \cosh 2\delta z - \sinh 2\delta z}{\sinh^3 2\delta z}. \tag{B.12}$$

Note that $F(0) = BA_s^2/3\delta$ and we consider $z = 0$ as the point at which the two pulses coincide. The point z_0 at which the two pulses are initially injected into the fiber is given by $z_0 = (t_v - t_u)/2\delta$, where t_u and t_v are the peak locations of the pulses at the injection point. Since F is a rapidly decaying function, much of the frequency shift occurs in the first one or two walk-off lengths l_{wo} of the fiber (we define $l_{\text{wo}} = c\tau/\Delta n$ which is given in normalized units by $l_{\text{wo}}/Z_0 = 1.763/\pi\delta$, Z_0 being the soliton period).

Suppose that after propagating a length l_1 the pulses undergo a discrete loss or gain, after which they continue to propagate a long length l_2. The transmission coefficient Γ at the discrete point is $\Gamma > 1$ for gain and $\Gamma < 1$ for loss. The overall frequency shift for the control pulse, if it propagates along the fast axis, is given by

$$\Delta\omega = F(l_1 + z_0) - F(z_0) + \Gamma[F(l_2 + l_1 + z_0) - F(l_1 + z_0)]$$
$$\simeq -F(z_0) + (1 - \Gamma) F(l_1 + z_0), \tag{B.13}$$

where we note that $F(l_2 + l_1 + z_0) \simeq F(\infty) = 0$. The resulting delay in the temporal location of the control pulse is given by $-\Delta\omega l_2$ (assuming $l_2 \gg l_1$) so that

$$\Delta\tilde{T} \simeq l_2[F(z_0) - (1 - \Gamma) F(l_1 + z_0)]. \tag{B.14}$$

Normalizing the change in time relative to the full width at half maximum pulse width τ, we find that $\Delta T/\tau = \Delta\tilde{T}/1.763$.

To illustrate the use of these formulas, consider the specific case of a $2l_{\text{wo}}$ length of birefringent fiber followed by a discrete loss with $\Gamma = 0.75$ and then coupled into a $35Z_0$ length of the same birefringent fiber. In Fig. B.3 we plot the shift in the control pulse ($\Delta T/\tau$)

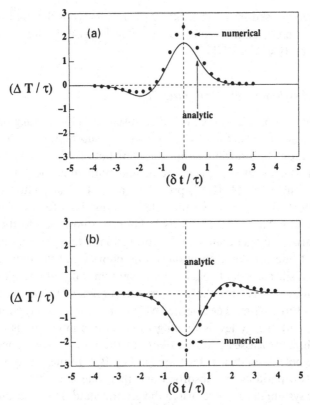

Fig. B.3 Time shift of the control pulse ($\Delta T/\tau$) versus the initial separation between the two pulses ($\delta t/\tau$) with the control pulse along the (a) fast and (b) slow axes. The solid curves are calculated from the analytic formulas (B.12–B.14), while the points are numerically computed from the coupled nonlinear Schrödinger equations.

versus the initial separation between the two pulses ($\delta t/\tau$) with the control along the fast and slow axes. The solid curves are calculated from Eqs. (B.12–B.14) for the choice of parameters: $\delta = 0.894$, $A_s = 0.48$, $A_c = 1.11$ and $B = \frac{2}{3}$. The peak around $\delta t = 0$ corresponds to the overlap of the two pulses at the input to the first fiber. The second peak occurs when the two pulses coincide at the discrete loss point, and the amplitude of the shift is proportional to the magnitude of the loss. As shown in Figs. 4.5 (c) and 4.5 (d), the frequency and time shift change signs in this case. To test the accuracy of the simple chirp picture of Eqs. (B.12–B.14), we also numerically solved the CNLS of Eqs. (B.3, B.4) using a split-step Fourier transform technique. The computed exact solutions are given by the dotted points in Fig. B.3, and we find a maximum discrepancy of 27% for the parameters used here. The difference results from the creation of "shadows" and radiative components, which arise because of the nonintegrability of the CNLS.

Cross-Raman effects in birefringent fibers

Since subpicosecond pulses are used in soliton-dragging and -trapping experiments, the Raman effect can substantially affect the pulse evolution. In birefringent optical fibers both the parallel and perpendicular Raman effects play a role, and both must be taken into account. Menyuk, et al. [B.8], provide a nice physical picture of the Raman effect and also derive the relevant equations for birefringent fibers. The Raman and Kerr effects in fibers are both due to the nonlinear polarizability generated by the absorption of two photons and the emission of one (or the absorption of one photon and the emission of two). Fused silica optical fibers have no quantum states at energies near $\hbar\omega_0$ or $2\hbar\omega_0$, at least for the telecommunications wavelengths in the range of 1.3 to 1.6 µm. The energy levels from nuclear interactions are far lower, and energy levels relating to electronic interactions are far higher. Three processes that contribute to the positive frequency portion of the polarizability are illustrated in Fig. B.4. In the first type (Fig. B.4(a)), two photons are absorbed and then one is emitted, and the molecular system does not return to the ground state. In the second and third processes (Figs. B.4(b), (c)) there is one absorption and one emission in any order, followed by another absorption, and the molecular system returns near the ground state before the last absorption. There are also three conjugate processes (not shown) with two emissions and one absorption, which contribute to the negative frequency portion of the polarizability. From the uncertainty principle, the lifetimes of the virtual transitions in Fig. B.4 are limited to the inverse of the frequency separation from the nuclear levels. Therefore, the process of Fig. B.4(a) is necessarily instantaneous, while the processes in Figs. B.4(b), (c) can have a finite delay time in the middle. The Raman effect is associated with processes with a finite time delay.

The CNLS can be modified to account for the Raman effect as [B.8]

$$-i\left(\frac{\partial u}{\partial z} + \delta\frac{\partial u}{\partial t}\right) = \frac{1}{2}\frac{\partial^2 u}{\partial t^2} + |u|^2 u + \tfrac{2}{3}|v|^2 u + i\frac{\partial u}{\partial z}\bigg|_{\text{Raman}}, \quad (B.15)$$

$$-i\left(\frac{\partial v}{\partial z} - \delta\frac{\partial v}{\partial t}\right) = \frac{1}{2}\frac{\partial^2 v}{\partial t^2} + |v|^2 v + \tfrac{2}{3}|u|^2 v + i\frac{\partial v}{\partial z}\bigg|_{\text{Raman}}. \quad (B.16)$$

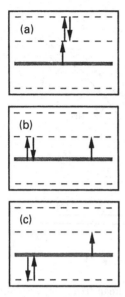

Fig. B.4 Schematic diagram of the molecular transitions that create the positive frequency component of the polarizability in the Kerr and Raman effects. The virtual levels at $-\hbar\omega_0$, $\hbar\omega_0$ and $2\hbar\omega_0$ are shown as dashed lines, while the nuclear levels are shown as solid lines. The electronic levels are not included. In process (a), where the system never returns to the nuclear levels, the whole process is essentially instantaneous. By contrast, in processes (b) and (c), where the system returns to the nuclear levels, there can be a finite time delay [B.8].

The Raman term in a birefringent fiber can be written in the most general form as [B.8]

$$
\begin{aligned}
\mathrm{i}\frac{\partial u(t)}{\partial z}\bigg|_{\mathrm{Raman}} = \; & u(t) \int_{-\infty}^{t} f_1(t-t')|u(t')|^2 \, \mathrm{d}t' \\
& + u(t) \int_{-\infty}^{t} f_2(t-t')|v(t')|^2 \, \mathrm{d}t' \\
& + v(t) \int_{-\infty}^{t} f_3(t-t')u(t')v^*(t') \, \mathrm{d}t' \\
& + v(t) \int_{-\infty}^{t} f_4(t-t')u^*(t')v(t')\mathrm{e}^{\mathrm{i}R\delta z} \, \mathrm{d}t',
\end{aligned}
$$

$$(B.17)$$

with a similar result for $\mathrm{i}\,\partial v/\partial z\,|_{\mathrm{Raman}}$. If the nonlinearity is assumed

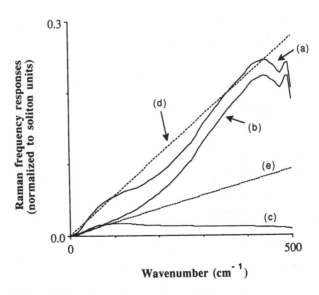

Fig. B.5 Parallel and perpendicular Raman frequency response functions and linear fits to their low-frequency portions. (a) $\text{Im}(\tilde{f}_1)$ from polarized Raman scattering; (b) $\text{Im}(\tilde{f}_2)$; (c) $\text{Im}(\tilde{f}_3)$ from depolarized Raman scattering; (d) linear fit to $\text{Im}(\tilde{f}_1)$; and (e) linear fit to $\text{Im}(\tilde{f}_2)$ and $\text{Im}(\tilde{f}_3)$ [B.9].

to be isotropic, then $f_1 = f_2 + f_3 + f_4$. Furthermore, we find that $f_3 = f_4$ since real polarizability depends only on the real fields. We note that the parallel Raman gain is contributed by $f_1(t)$, while the perpendicular Raman gain is contributed by $f_3(t)$. The quantity $f_2(t)$ contributes a frequency shift but no gain.

Experimental measurements usually yield the Fourier transforms of the time-domain response functions [B.9], which are defined as

$$\tilde{f}_n = \int_{-\infty}^{\infty} f_n(t) e^{i\Omega t} \, dt \qquad (n = 1, 2, 3). \tag{B.18}$$

By measuring the self- and cross-polarized Raman gains as a function of frequency using a cw source, we can determine $\text{Im}(\tilde{f}_1)$ and $\text{Im}(\tilde{f}_3)$ [B.10]. We can then infer the functions f_1 and f_3 since the real parts of $\tilde{f}_1$ and $\tilde{f}_3$ are related by the Kramers–Kronig relations to the measured imaginary parts. Finally, the relation given above that $f_1 = f_2 + 2f_3$ can be used to obtain f_2. As an example, Fig. B.5 shows estimates of the three response functions in fused silica fibers and linear fits to the low-frequency portions [B.9].

When the pulse durations are sufficiently larger (e.g., $\tau > 100\,\text{fs}$), then the response functions can be Taylor expanded and replaced by the first linear terms. The approximate Raman contributions in Eq. (B.17) then become

$$\text{i}\frac{\partial u}{\partial z}\bigg|_{\text{Raman}} = -\left(c_1 u\frac{\partial|u|^2}{\partial t} + c_2 u\frac{\partial|v|^2}{\partial t} + c_3 v\frac{\partial uv^*}{\partial t}\right) \qquad \text{(B.19)}$$

and

$$\text{i}\frac{\partial v}{\partial z}\bigg|_{\text{Raman}} = -\left(c_1 v\frac{\partial|v|^2}{\partial t} + c_2 v\frac{\partial|u|^2}{\partial t} + c_3 u\frac{\partial u^*v}{\partial t}\right), \qquad \text{(B.20)}$$

where c_1, c_2 and c_3 are constant coefficients and $c_1 = c_2 + 2c_3$. For example, from the straight-line approximations in Fig. B.5 (curves (d) and (e)) and assuming $300\,\text{fs}$ pulses, the values for the coefficients are $c_1 \simeq 0.0174$ and $c_2 = c_3 = \tfrac{1}{3}c_1$.

Modulational instability between orthogonal axes

The final topic in this Appendix is a four-wave-mixing phenomena owing to the birefringence in a bimodal fiber. For the standard nonlinear Schrödinger equation of Eq. (A.11), modulational instability gain occurs only in the anomalous group-velocity dispersion regime. However, when we add birefringence to the phase-matching condition there can also be cross-modulational instability in the normal-dispersion regime [B.11, B.12]. To obtain the gain formulas, we include in the CNLS terms corresponding to phase- and group-velocity mismatch and the coherence terms, which are responsible for phase conjugation and polarization instability. Unlike Eqs. (B.1) and (B.2) where the phase mismatch is included in the coherence terms, here we explicitly include a separate term. Let us assume that the pump wave is along u and the probe or signal is along the orthogonal v axis ($|v| \ll |u|$). The generalized CNLS can be written as

$$-\text{i}\left(\frac{\partial u}{\partial z} + 2\delta\frac{\partial u}{\partial t}\right) = \tfrac{1}{2}\beta\frac{\partial^2 u}{\partial t^2} + |u|^2 u + \tfrac{2}{3}|v|^2 u + \tfrac{1}{3}v^2 u^* + \alpha u,$$

$$\text{(B.21)}$$

$$-\text{i}\frac{\partial v}{\partial z} = \tfrac{1}{2}\beta\frac{\partial^2 v}{\partial t^2} + |v|^2 v + \tfrac{2}{3}|u|^2 v + \tfrac{1}{3}u^2 v^*, \qquad \text{(B.22)}$$

where 2δ is the group-velocity mismatch, $\alpha = \Delta k z_c = 2\pi\,\Delta n\,z_c/\lambda$ is the phase-velocity mismatch, and the normalized dispersion is β ($\beta > 0$ for anomalous dispersion and $\beta < 0$ for normal dispersion). For simplicity we can collect the entire phase- and group-velocity mismatch in Eq. (B.21) for the quasi-cw pump u.

If we retain terms only to first order in $|v|$, then Eqs. (B.21) and (B.22) reduce to

$$-i\left(\frac{\partial u}{\partial z} + 2\delta\frac{\partial u}{\partial t}\right) = \tfrac{1}{2}\beta\frac{\partial^2 u}{\partial t^2} + |u|^2 u + \alpha u, \tag{B.23}$$

$$-i\frac{\partial v}{\partial z} = \tfrac{1}{2}\beta\frac{\partial^2 v}{\partial t^2} + \tfrac{2}{3}|u|^2 v + \tfrac{1}{3}u^2 v^*. \tag{B.24}$$

We assume cw solutions of the form

$$u = a \exp\{i(a^2+\alpha)z\}, \tag{B.25}$$

$$v = V(z,t)\exp\{i(a^2+\alpha)z\}, \tag{B.26}$$

where a is real and the pump in relation (B.25) is a solution to Eq. (B.23). Introducing (B.26) into relation (B.24), we obtain the master equation

$$-i\frac{\partial V}{\partial z} = \tfrac{1}{2}\beta\frac{\partial^2 V}{\partial t^2} - \alpha V - \tfrac{1}{3}a^2 V + \tfrac{1}{3}a^2 V^*, \tag{B.27}$$

where the phase-velocity mismatch enters, but the group-velocity mismatch drops out because of the assumed cw solutions.

As in the derivation in Appendix A, we assume a perturbing field that is frequency shifted by $\pm\Omega$ from the pump

$$V(z,t) = A(z)e^{-i\Omega t} + B^*(z)e^{+i\Omega t}. \tag{B.28}$$

Introducing (B.28) into relation (B.27) we obtain the equations

$$\left(\tfrac{1}{3}a^2 + \alpha + \tfrac{1}{2}\beta\Omega^2 - i\frac{\partial}{\partial z}\right)A = \tfrac{1}{3}a^2 B; \tag{B.29(a)}$$

$$\left(\tfrac{1}{3}a^2 + \alpha + \tfrac{1}{2}\beta\Omega^2 - i\frac{\partial}{\partial z}\right)B^* = \tfrac{1}{3}a^2 A^*. \tag{B.29(b)}$$

If we assume an exponentially growing amplitude $A \sim e^{\kappa z}$, then combining Eqs. (B.29) we obtain the dispersion relation

$$\kappa^2 = (\tfrac{1}{3}a^2)^2 - (\tfrac{1}{3}a^2 + \alpha + \tfrac{1}{2}\beta\Omega^2)^2. \tag{B.30}$$

The maximum gain coefficient $\kappa_{max} = \tfrac{1}{3}a^2$ is realized when

$$\tfrac{1}{3}a^2 + \alpha + \tfrac{1}{2}\beta\Omega^2 = 0, \tag{B.31}$$

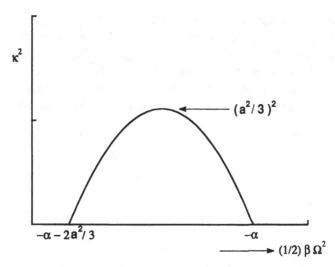

Fig. B.6 Dispersion relation for modulational-instability gain between orthogonal axes in a birefringent fiber.

which, therefore, corresponds to an effective phase matching between the pump and sidebands. For example, $\alpha + \frac{1}{2}\beta\Omega^2 = 0$ is the phase-matching condition at low power ($a \to 0$). The gain coefficient goes to zero when

$$\tfrac{1}{3}a^2 + \alpha + \tfrac{1}{2}\beta\Omega^2 = \pm \tfrac{1}{3}a^2. \tag{B.32}$$

Thus, we obtain net gain when $\frac{1}{2}\beta\Omega^2$ is within the interval $(-\alpha - \frac{2}{3}a^2, -\alpha)$, which is shown schematically in Fig. B.6.

Let us consider three limiting cases for the dispersion relation in Eq. (B.30). First, when the pump field is zero $a = 0$, then

$$\kappa^2 = -(\alpha + \tfrac{1}{2}\beta\Omega^2)^2 > 0 \tag{B.33}$$

and there is no gain for real values of α and β. Second, when operating in the anomalous dispersion ($\beta = |\beta|$), there is a finite gain window only for $\alpha < 0$; i.e., the pump wave must be along the fast axis, which has the lower index. The bandwidth in the anomalous dispersion is given by

$$|\alpha| - \tfrac{2}{3}a^2 \leqslant \tfrac{1}{2}|\beta|\Omega^2 \leqslant |\alpha|. \tag{B.34}$$

Finally, for normal dispersion ($\beta = -|\beta|$) the gain bandwidth becomes

$$\alpha \leqslant \tfrac{1}{2}|\beta|\Omega^2 \leqslant \alpha + \tfrac{2}{3}a^2 \quad \text{and} \quad \tfrac{1}{2}|\beta|\Omega^2 > 0. \tag{B.35}$$

If $\alpha > 0$, then the birefringence "biases-up" the pump and shifts the gain-frequency range to higher frequencies. If $\alpha = 0$ then frequencies between $0 \leq |\Omega| \leq 2|a|/\sqrt{3|\beta|}$ receive gain. Finally, when $\alpha < 0$ a finite gain window exists only for $|\alpha| \leq \frac{2}{3}a^2$. Therefore, adding birefringence to the phase-matching condition does produce modulational-instability gain even in the normal-dispersion regime.

Interaction between parallel polarized solitons

When solitons of the nonlinear Schrödinger equation (NLSE) in a common state of polarization pass through one another, their amplitudes and frequencies are unchanged, and there is no residual scattering into sideband frequencies. The sole residual results of their nonlinear interaction are that they are displaced, pushed farther apart than they would otherwise have been, and that they acquire phase shifts. Remarkably, the net displacements and phase shifts *after* the pulses separate do not depend on the relative phases of the solitons during their collisions. Furthermore, the displacements are additive when, for example, collisions occur with three or more solitons. On the other hand, the state of a pair of solitons *during* a collision can depend strongly on their relative phase, particularly if their amplitudes and velocities are nearly the same.

The solution of the NLSE for an arbitrary number of solitons was first obtained by Zakharov and Shabat [C.1]. As discussed in Appendix A, for the normalized NLSE

$$-i\frac{\partial u}{\partial z} = \frac{1}{2}\frac{\partial^2 u}{\partial t^2} + u^* u^2, \tag{C.1}$$

single-soliton solutions have the general form

$$u = A \operatorname{sech}(At - q) \exp\{-i(\Omega t + \phi)\}. \tag{C.2}$$

Here the amplitude A and frequency Ω are constants, $dq/dz = -A\Omega$ describes the motion of the center of the soliton, and $d\phi/dz = \frac{1}{2}(A^2 - \Omega^2)$ describes the motion of its phase. The general N soliton solution can be obtained from the solution of N algebraic equations [C.2]. When solitons are well separated, each one takes the form of Eq. (C.2), with an added displacement and phase shift that reflects how many of the other solitons are at earlier or later times. The displacements and phase shifts change as the solitons pass through one another as a result of this interaction.

The general two-soliton function can be written in the form [C.2],

$$u = \frac{A_1 e^{i\theta_1}(\beta^*\rho e^{\xi_2} + \beta\rho^* e^{-\xi_2}) - A_2 e^{i\theta_2}(\beta\rho e^{\xi_1} + \beta^*\rho^* e^{-\xi_1})}{|\rho|^2 \cosh(\xi_1 + \xi_2) + |\beta|^2 \cosh(\xi_1 - \xi_2) - A_1 A_2 \cos(\theta_1 - \theta_2)},$$
(C.3)

where $2\beta = A_1 + A_2 - i(\Omega_1 - \Omega_2)$ and $2\rho = A_1 - A_2 - i(\Omega_1 - \Omega_2)$, while, for $j = 1$ or 2, $\xi_j = A_j t - q_j$ and $\theta_j = -\Omega_j t + \phi_j$. As in the single-soliton case, the amplitudes A_j and frequencies Ω_j are constants, and the soliton motions are described by $dq_j/dz = -A_j\Omega_j$ and $d\phi_j/dz = \frac{1}{2}(A_j^2 - \Omega_j^2)$. Note that the two terms of the numerator of Eq. (C.3) are related by the interchange of indices, which does not change the denominator. If either A_1 or A_2 is zero, then Eq. (C.3) reduces to the single-soliton form of Eq. (C.2).

If $\Omega_1 \neq \Omega_2$, the solitons are unbound. If they collide near $z = 0$, they are always well separated for sufficiently large negative or positive values of z. When they are well separated, then near soliton #1, where ξ_1 is small, either $e^{\xi_2} \gg 1$ (soliton #2 at a much earlier time), or $e^{-\xi_2} \gg 1$ (soliton #2 at a much later time). In the former case, for example, we need only keep the e^{ξ_2} terms of Eq. (C.3), yielding

$$u = \frac{2A_1 e^{i\theta_1}}{(\rho/\beta)^* e^{\xi_1} + (\beta/\rho)e^{-\xi_1}}.$$
(C.4)

Then if we define the quantities δ and ψ_1 from

$$\exp(\delta - i\psi_1) \equiv \frac{\beta}{\rho} = \frac{A_1 + A_2 - i(\Omega_1 - \Omega_2)}{A_1 - A_2 - i(\Omega_1 - \Omega_2)},$$
(C.5)

we get

$$u = A_1 \operatorname{sech}(\xi_1 - \delta)\exp\{i(\theta_1 + \psi_1)\}.$$

Thus, the distant presence of soliton #2 causes a displacement in time of soliton #1 by an amount δ/A_1 and a phase shift ψ_1. Note that the displacement of soliton #1 is away from soliton #2, since δ is always positive. In the latter case (soliton #2 at a later time), the displacement and phase shift have the same magnitudes but opposite signs. Similar results for soliton #2 are obtained by interchanging the indices. We see that the two-soliton function, Eq. (C.3), contains the kinetics of the collision of the two solitons, and the displacements and phase shifts change sign during the collision. The total displacement in time of soliton #1 in the collision is

$$\frac{2\delta}{A_1} = \frac{1}{A_1} \ln \frac{(A_1 + A_2)^2 + (\Omega_1 - \Omega_2)^2}{(A_1 - A_2)^2 + (\Omega_1 - \Omega_2)^2}. \tag{C.6}$$

We find from Eq. (C.5) that neither the displacements nor phase shifts depend on the relative phase of the soliton pair when they collide. However, during the collision the two solitons exist in the region where ξ_1 and ξ_2 are both small, and here their relative phase $\theta_1 - \theta_2$ becomes important. For example, at the central point of the collision, where z and t are determined by $\xi_1 = \xi_2 = 0$, the field can be expressed as

$$u = e^{i\bar{\theta}} \frac{(A_1 + A_2)|\rho|^2 \cos \Phi + i(A_1 - A_2)|\beta|^2 \sin \Phi}{|\rho \cos \Phi|^2 + |\beta \sin \Phi|^2}, \tag{C.7}$$

where $\bar{\theta} = \frac{1}{2}(\theta_1 + \theta_2)$ and $\Phi = \frac{1}{2}(\theta_1 - \theta_2)$. The power $|u|^2$ at this point varies between $(A_1 + A_2)^2$ for the in-phase case $\theta_2 = \theta_1$, and $(A_1 - A_2)^2$ for the out-of-phase case $\theta_2 = \theta_1 + \pi$. The difference is quite striking when the two solitons have nearly the same amplitudes and frequencies, as illustrated in Figs. 3.14 and C.1. In Fig. 3.14 we show solutions for different phases and $-40 \leqslant z \leqslant 40$, so the solitons are well separated before and after the collision. In the in-phase case the two solitons appear to attract each other and cross, while in the out-of-phase case they seem to repel each other and stay apart. In the latter case, however, the amplitudes are exchanged. Therefore, the solitons always emerge from the collision with the same amplitudes and frequencies with which they entered and with phase-independent displacements in time.

Figure C.1 concentrates more on the interaction region between $-10 \leqslant z \leqslant 10$, and we see dramatically different behavior for different phases. The dashed lines show the phase independent asymptotic incoming and outgoing trajectories of the peak (or centroid) of the slower soliton. We see that the interaction displaces each soliton away from the other. Although the two solitons have the same initial amplitude, note that the distribution between the two pulses changes with phase. For example, for a phase difference of $\frac{1}{32}\pi$ and $\frac{1}{16}\pi$ the pulse traveling to the right at $z = 0$ is much taller than the pulse traveling to the left. Also, for phase differences less than $\frac{1}{4}\pi$ the pulse peaks remain inside the asymptotic lines, while for phase differences larger than $\frac{1}{4}\pi$ the pulse peaks stay outside the lines. No Raman effect is assumed in these solutions, so the center of mass of the pulse pair always remains at $t = 0$. If the phase difference is of the opposite sign, then the pulses simply interchange roles.

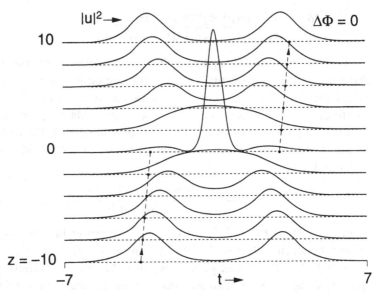

(a) The phase difference at the center is 0.

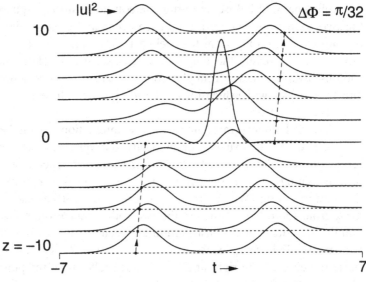

(b) The phase difference at the center is $\frac{1}{32}\pi$.

Fig. C.1 Collision of two solitons with with initially equal amplitudes ($A_j = 1$). The frequencies of the two solitons are $\Omega = \pm 0.05$
and the collision center is at $t = z = 0$.

(Continued on the opposite page.)

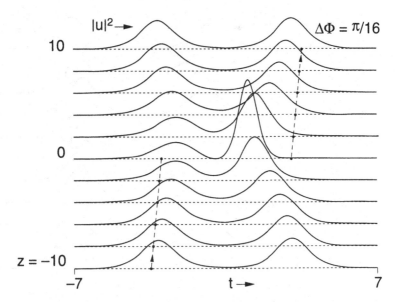

(c) The phase difference at the center is $\frac{1}{16}\pi$.

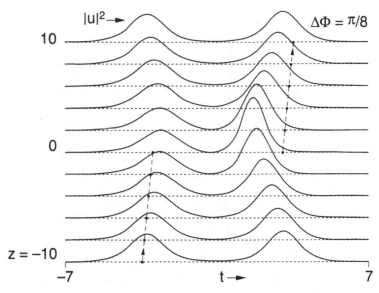

(d) The phase difference at the center is $\frac{1}{8}\pi$.

Fig. C.1 (*continued*) The dashed lines are the asymptotic incoming and outgoing trajectories of the slower soliton, showing its time displacement because of the collision. The phase difference at the center ($t = z = 0$) in each of the cases (a)–(f) is given under the figure.

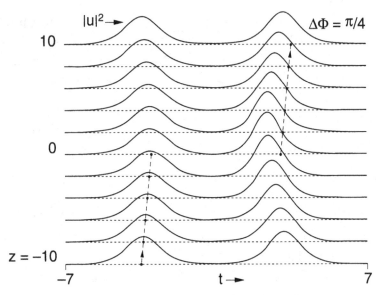

(e) The phase difference at the center is $\frac{1}{4}\pi$.

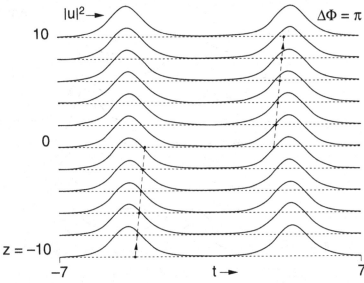

(f) The phase difference at the center is π.

Fig. C.1 (*continued*)

Raman gain in soliton collisions

The elastic collision property of solitons is violated when Raman effects or soliton self-frequency shift are included. Because of Raman gain, the lower-frequency pulse is amplified while the higher-frequency pulse is attenuated. Consider the NLSE including Raman effects

$$-i\frac{\partial u}{\partial z} = \frac{1}{2}\frac{\partial^2 u}{\partial t^2} + u \int_{-\infty}^{\infty} ds \, f(s)|u(t-s)|^2, \qquad (C.8)$$

where the response function $f(t)$ is a real delay function satisfying

$$f(-|t|) = 0 \quad \text{and} \quad \int_0^{\infty} dt \, f(t) = 1.$$

Stolen, et al. [C.3], have calculated the response function for pure silica-core fibers, which is included in Fig. C.2. Some of the details of the response function will vary from fiber to fiber because of differences in dopants and drawing conditions, but the general features are consistent. Figure C.2 shows that the response function is basically a decaying sinusoidal oscillation. The oscillation period corresponds to the frequency of the peak of the Raman-gain spectrum, and the decay rate corresponds to the width of the gain spectrum (cf. Fig. A.7). Additional peaks in the gain spectrum will result in a mixture and interference between sinusoides with different periods in the response function.

We separate the input into two parts $u = u_1 + u_2$ and only treat the simplest case when the frequency spectra of u_1 and u_2 are individually narrower than the separation between them. In this approximation the equations can be solved analytically. Considering only frequencies of the band corresponding to u_1, we get from Eq. (C.8) the expression [C.4]

$$\frac{\partial}{\partial z}|u_1|^2 = -\text{Im}\left[\frac{\partial}{\partial t}\left(u_1^*\frac{\partial u_1}{\partial t}\right) + 2u_1^* u_2 \int_0^{\infty} ds \, f(s)(u_2^* u_1)_{t-s}\right].$$
$$(C.9)$$

If we take the mean frequency of the u_1 band to be zero and that of the u_2 band to be $\bar{\Omega}_2$, we can approximate Eq. (C.9) over a short distance by

$$\frac{\partial}{\partial z}|u_1|^2 \cong -\text{Im}\left(2|u_1|^2|u_2|^2 \int_0^{\infty} ds \, f(s)e^{-i\bar{\Omega}_2 s}\right). \qquad (C.10)$$

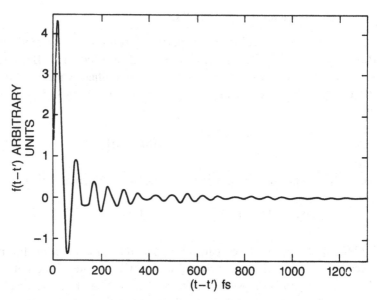

Fig. C.2 The Raman time response function for silica-core fibers
[C.3].

For $\bar{\Omega}_2$ well below the peak of the Raman gain spectrum, we can
approximate the imaginary part of the integral in (C.10) by expanding
the exponential to $-\bar{\Omega}_2 t_d$, where

$$t_d = \int_0^\infty dt \; tf(t).$$

Using the latter approximation, we have from relation (C.10)

$$\frac{\partial}{\partial z} \ln |u_1|^2 = 2|u_2|^2 \bar{\Omega}_2 t_d. \tag{C.11}$$

In the case when u_2 is a soliton, so that

$$|u_2|^2 = \text{sech}^2(t + \bar{\Omega}_2 z),$$

we can integrate Eq. (C.11) over the region of passage of the soliton
to give

$$\frac{|u_1|^2_{\text{final}}}{|u_1|^2_{\text{initial}}} = \exp\{4t_d \, \text{sgn}(\bar{\Omega}_2)\}, \tag{C.12}$$

where sgn is the sign of $\bar{\Omega}_2$ and equals either $+1$ or -1.

As expected, the soliton will gain energy in passing fields of higher
frequency ($\bar{\Omega}_2 > 0$) and lose energy in passing fields of lower frequency

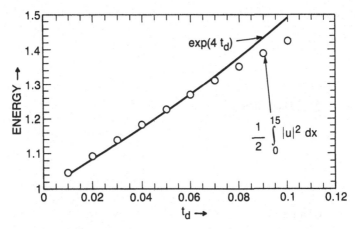

Fig. C.3 Energy increase for lower-frequency pulse after collision with a higher-frequency pulse. The solid curve corresponds to e^{4t_d}, and the circles to energies computed by solving Eq. (C.8) under the approximation of relation (A.35) with $L = 3Z_0$ and the initial condition $u = \operatorname{sech}(t+3.5)\exp(i2t) + \operatorname{sech}(t-3.5)\exp\{i(-2t+\tfrac{1}{2}\pi)\}$.

($\bar{\Omega}_2 < 0$). In our crude approximation, the energy transfer is independent of the relative velocity $|\bar{\Omega}_2|$ because the increased rate of energy transfer is balanced by the decreased interaction time. Note also that Eq. (C.12) is accurate only when $|u_2|^2$ does not change much; i.e., when the energy change of the soliton is small. To validate the result from Eq. (C.12), we compute the energy exchange in the passage of two solitons, with the results shown in Fig. C.3. One can see that Eq. (C.12) is not a bad approximation. At smaller relative velocities, the energy exchange is somewhat greater than predicted by Eq. (C.12). In such a case the condition of nonoverlapping spectra does not hold so well.

References

Chapter 1

(1) D. A. B. Miller, "Device requirements for digital optical processing" in *Digital Optical Computing*, ed. R. A. Athale, SPIE Critical Reviews of Optical Science and Technology, **CR35**, 68 (1990).

(2) P. W. Smith, Bell System Tech. Jour. **61**, 1975 (1982).

(3) A. Lattes, H. A. Haus, F. J. Leonberger and E. P. Ippen, IEEE J. Quantum Electron. **QE-19**, 1718 (1983).

(4) V. Mizrahi, K. W. DeLong, G. I. Stegeman, M. A. Saifi and M. J. Andrejco, Opt. Lett. **14**, 1140 (1989).

(5) S. T. Ho, C. E. Soccolich, M. N. Islam, W. S. Hobson, A. F. J. Levi and R. E. Slusher, Appl. Phys. Lett. **59**, 2558 (1991).

(6) M. N. Islam, C. E. Soccolich, R. E. Slusher, A. F. J. Levi, W. S. Hobson and M. G. Young, J. Appl. Phys. **71** 1927 (1992).

(7) J. S. Aitchison, A. H. Kean, C. N. Ironside, A. Villeneuve and G. I. Stegeman, Electron. Lett. **27**, 1709 (1991).

(8) G. I. Stegeman and R. H. Stolen, J. Opt. Soc. Am. B. **6**, 652 (1989).

(9) L. F. Mollenauer, J. P. Gordon and S. G. Evangelides, Laser Focus World, Nov. 1991, pp. 159–70.

(10) N. J. Doran and D. Wood, Opt. Lett. **13**, 56 (1988).

(11) C. R. Menyuk, J. Opt. Soc. Am. B. **5**, 392 (1988).

(12) J. Satsuma and N. Yajima, Prog. Theor. Phys. Suppl. **55**, 284 (1974).

Chapter 2

(1) M. A. Duguay and J. W. Hansen, Appl. Phys. Lett. **15**, 192 (1969).

(2) R. H. Stolen, J. Botineau and A. Ashkin, Opt. Lett. **7**, 512 (1982); R. H. Stolen and A. Ashkin, Appl. Phys. Lett. **22**, 294 (1973).

(3) N. J. Halas, D. Krokel and D. Grischkowsky, Appl. Phys. Lett. **50**, 886 (1987).

(4) T. Morioka, M. Saruwatari and A. Takeda, Electron. Lett. **23**, 453 (1987); T. Morioka and M. Saruwatari, IEEE J. Select. Areas Commun. **6**, 1186 (1988).

(5) M. N. Islam, C. E. Soccolich, J. P. Gordon and U. C. Paek, Opt. Lett. **15**, 21 (1990).

(6) L. F. Mollenauer, R. H. Stolen, J. P. Gordon and W. J. Tomlinson, Opt. Lett. **8**, 289 (1983).

(7) M. N. Islam, E. R. Sunderman, C. E. Soccolich, I. Bar-Joseph, N. Sauer, T. Y. Chang and B. I. Miller, IEEE J. Quantum Electron. **25**, 2454 (1989).

(8) C. R. Menyuk, J. Opt. Soc. Am. B. **5**, 392 (1988).

(9) R. H. Stolen and J. E. Bjorkholm, IEEE J. Quantum Electron. **QE-18**, 1062 (1982).

(10) P. A. Andrekson, N. A. Olsson, J. R. Simpson, T. Tanbun-Ek, R. A. Logan and M. Haner, Electron. Lett. **27**, 922 (1991).

(11) M. N. Islam, S. P. Dijaili and J. P. Gordon, Opt. Lett. **13**, 518 (1988).

(12) J. P. Gordon, Opt. Lett. **11**, 662 (1986).

(13) F. M. Mitschke and L. F. Mollenauer, Opt. Lett. **11**, 659 (1986).

(14) C. E. Soccolich and M. N. Islam, Opt. Lett. **14**, 645 (1989).

(15) S. M. Jensen, IEEE J. Quantum Electron. **QE-18**, 1580 (1982).

(16) S. R. Friberg, A. M. Weiner, Y. Silberberg, G. Sfez and P. W. Smith, Opt. Lett. **13**, 904 (1988).

(17) S. Trillo, S. Wabnitz, R. H. Stolen, G. Assanto, C. T. Seaton and G. I. Stegeman, Appl. Phys. Lett. **49**, 1224 (1986).

(18) S. Trillo, S. Wabnitz, N. Finlayson, W. C. Banyai, C. T. Seaton and G. I. Stegeman, Appl. Phys. Lett. **53**, 837 (1988).

(19) A. M. Weiner, Y. Silberberg, H. Fouckhardt, D. E. Leaird, M. A. Saifi, M. J. Andrejco and P. W. Smith, IEEE J. Quantum Electron. **25**, 2648 (1989).

(20) S. Trillo, S. Wabnitz, E. M. Wright and G. I. Stegeman, Opt. Lett. **13**, 672 (1988).

(21) M. J. LaGasse, D. Liu-Wong, J. G. Fujimoto and H. A. Haus, Opt. Lett. **14**, 311 (1989).

(22) N. J. Doran and D. Wood, Opt. Lett. **13**, 56 (1988).

(23) K. J. Blow, N. J. Doran and B. K. Nayar, Opt. Lett. **14**, 754 (1989).

(24) M. N. Islam, E. R. Sunderman, R. H. Stolen, W. Pleibel and J. R. Simpson, Opt. Lett. **14**, 811 (1989).

(25) H. A. Haus and M. N. Islam, IEEE J. Quantum Electron. **QE-21**, 1172 (1985).

(26) A. Lattes, H. A. Haus, F. J. Leonberger and E. P. Ippen, IEEE J. Quantum Electron. **QE-19**, 1718 (1983).

(27) K. J. Blow, N. J. Doran, B. K. Nayar and B. P. Nelson, Opt. Lett. **15**, 248 (1990).

(28) J. D. Moores, K. Bergman, H. A. Haus and E. P. Ippen, Opt. Lett. **16**, 138 (1991).

(29) H. Avramopoulos, P. M. W. French, M. C. Gabriel and N. A. Whitaker, IEEE Phot. Tech. Lett. **3**, 235 (1991).

(30) T. J. Cloonan, M. Herron, F. A. P. Tooley, G. W. Richards, F. B. McCormick, E. Kerbis, J. L. Brubaker and A. L. Lentine, IEEE Phot. Tech. Lett. **2**, 438 (1990).

(31) G. W. Richards, "Extended Generalized Shuffle (EGS) Network: Advantageous Attributes with respect to Free Space Photonic Switching" (talk presented at 1989 Photonic Switching Symposium, AT&T Bell Laboratories, Holmdel, N.J.).
(32) G. I. Stegeman and R. H. Stolen, J. Opt. Soc. Am. B. **6**, 652 (1989)

Chapter 3

(1) D. A. B. Miller, "Device requirements for digital optical processing" in *Digital Optical Computing*, ed. R. A. Athale, SPIE Critical Reviews of Optical Science and Technology, **CR35**, 68 (1990).
(2) R. W. Keyes, Reviews of Modern Physics, **61**, 279 (1989).
(3) R. W. Keyes, *The Physics of VLSI Systems*, New York: Addison-Wesley Publishing Co., 1987.
(4) M. N. Islam, C. E. Soccolich and D. A. B. Miller, Opt. Lett. **15**, 909 (1990); M. N. Islam, Opt. Lett. **15**, 417 (1990).
(5) J. P. Gordon, Opt. Lett. **11**, 662 (1986).
(6) M. N. Islam, L. F. Mollenauer, R. H. Stolen, J. R. Simpson and H. T. Shang, Opt. Lett. **12**, 625 (1987).
(7) M. N. Islam, C. R. Menyuk, C.-J. Chen and C. E. Soccolich, Opt. Lett. **16**, 214 (1991).
(8) M. N. Islam, E. R. Sunderman, C. E. Soccolich, I. Bar-Joseph, N. Sauer, T. Y. Chang and B. I. Miller, IEEE J. Quantum Electron. **25**, 2454 (1989).
(9) M. N. Islam, C. D. Poole and J. P. Gordon, Opt. Lett. **14**, 1011 (1989).
(10) M. N. Islam, C.-J. Chen and C. E. Soccolich, Opt. Lett. **16**, 593 (1991).
(11) M. N. Islam, C. E. Soccolich, C.-J. Chen, K. S. Kim, J. R. Simpson and U. C. Paek, Electron. Lett. **27**, 130 (1991).
(12) M. N. Islam, C. E. Soccolich, J. P. Gordon and U. C. Paek, Opt. Lett. **15**, 21 (1990).
(13) M. N. Islam, C. E. Soccolich, R. E. Slusher, A. F. J. Levi, W. S. Hobson and M. G. Young, J. Appl. Phys. **71** 1927 (1992).
(14) S. T. Ho, C. E. Soccolich, M. N. Islam, W. S. Hobson, A. F. J. Levi and R. E. Slusher, Appl. Phys. Lett. **59**, 2558 (1991).
(15) M. N. Islam, C. E. Soccolich, R. E. Slusher, W. S. Hobson and A. F. J. Levi, "Nonlinearity Near Half-Gap in Bulk and Quantum Well GaAs/AlGaAs Waveguides" (to be published in *Proc. of the VIIth International Symposium on Ultrafast Processes in Spectroscopy*, ed. A. Laubereau (Bristol, England: Adam Hilger)).
(16) E. Fredkin and T. Toffoli, Int. J. Theor. Phys. **21**, 219 (1982).
(17) M. N. Islam and C. E. Soccolich, Opt. Lett. **16**, 1490 (1991).
(18) J. P. Gordon, Opt. Lett. **8**, 596 (1983).
(19) V. E. Zahkarov and A. B. Shabat, Zh. Eksp. Teor. Fiz. **61**, 118 (1971); Sov. Phys. JETP **34**, 62 (1972).
(20) C. R. Menyuk, Opt. Lett. **12**, 614 (1987).
(21) C. R. Menyuk, J. Opt. Soc. Am. B. **5**, 392 (1988).

(22) M. N. Islam, Opt. Lett. **14**, 1257 (1989).
(23) M. W. Chbat, B. J. Hong, M. N. Islam, C. E. Soccolich, P. R. Prucnal and K. R. German, "Ultrafast Soliton Trapping AND-Gate" (submitted to J. Lightwave Tech.).
(24) C. E. Soccolich, M. N. Islam, M. W. Chbat and P. R. Prucnal, "Cascade of Ultrafast Soliton Dragging and Trapping Logic Gates" (submitted to IEEE Phot. Tech. Lett.).
(25) J. R. Sauer, M. N. Islam and S. P. Dijaili, "A Soliton Ring Network" (to be published in J. Lightwave Tech.).

Chapter 4

(1) J. R. Sauer, M. N. Islam and S. P. Dijaili, " A Soliton Ring Network" (to be published in J. Lightwave Tech.).
(2) S. P. Dijaili, J. S. Smith and A. Dienes, Appl. Phys. Lett. **55**, 418 (1989).
(3) M. N. Islam, C.-J. Chen and C. E. Soccolich, Opt. Lett. **16**, 484 (1991).
(4) M. N. Islam, C. E. Soccolich, C.-J. Chen, K. S. Kim, J. R. Simpson and U. C. Paek, Electron. Lett. **27**, 131 (1991).
(5) M. N. Islam, C. R. Menyuk, C.-J. Chen and C. E. Soccolich, Opt. Lett. **16**, 214 (1991).
(6) J. P. Gordon, Opt. Lett. **11**, 662 (1986).
(7) M. N. Islam, C. E. Soccolich and D. A. B. Miller, Opt. Lett. **15**, 909 (1990).
(8) D. B. Sarrazin, H. F. Jordon and V. P. Heuring, Appl. Opt. **29**, 627 (1990).
(9) M. N. Islam, C. E. Soccolich, S. T. Ho, R. E. Slusher, W. S. Hobson and A. F. J. Levi, Opt. Lett. **16**, 1116 (1991).
(10) S. T. Ho, C. E. Soccolich, M. N. Islam, W. S. Hobson, A. F. J. Levi and R. E. Slusher, Appl. Phys. Lett. **59**, 2558 (1991).
(11) M. N. Islam, C. E. Soccolich, R. E. Slusher, A. F. J. Levi, W. S. Hobson and M. G. Young, J. Appl. Phys. **71** 1927 (1992).
(12) M. N. Islam, C. E. Soccolich, R. E. Slusher, W. S. Hobson and A. F. J. Levi, "Nonlinearity Near Half-Gap in Bulk and Quantum Well GaAs/AlGaAs Waveguides" (to be published in *Proc. of the VIIth International Symposium on Ultrafast Processes in Spectroscopy*, ed. A. Laubereau (Bristol, England: Adam Hilger)).
(13) R. H. Stolen and C. Lin, Phys. Rev. A **17**, 1448 (1978).
(14) M. N. Islam, L. F. Mollenauer, R. H. Stolen, J. R. Simpson and H. T. Shang, Opt. Lett. **12**, 625 (1987).
(15) M. N. Islam, C. E. Soccolich, C.-J. Chen, U.-C. Paek, C. M. Schroeder, D. J. DiGiovanni and J. R. Simpson, Opt. Lett. **16**, 593 (1991).
(16) O. E. Martinez, IEEE J. Quantum Electron. **QE-23**, 59 (1987).
(17) J. Kuhl and J. Heppner, IEEE J. Quantum Electron. **QE-22**, 182 (1986).
(18) C. R. Menyuk, J. Opt. Soc. Am. B. **5**, 392 (1988).
(19) C. E. Soccolich, M. N. Islam, K. Möllmann, W. Gellermann and K. R. German, "Passively Modelocked Femtosecond Color Center Lasers in the Erbium Gain Band" (submitted to Appl. Phys. Lett.).

(20) E. Desurvire, C. R. Giles and J. R. Simpson, J. Lightwave Tech. **7**, 2095 (1989).

Chapter 5

(1) P. W. Smith, Bell System Tech. Jour. **61**, 1975 (1982); R. W. Keyes, *The Physics of VLSI Systems*, New York: Addison-Wesley Publishing Co., 1987.
(2) A. Lattes, H. A. Haus, F. J. Leonberger and E. P. Ippen, IEEE J. Quantum Electron. **QE-19**, 1718 (1983).
(3) L. F. Mollenauer, J. P. Gordon and S. G. Evangelides, Laser Focus World, Nov. 1991, pp. 159–70.
(4) S. K. Korotky and J. J. Veselka, "Efficient switching of a 72-Gbit/s Ti:LiNbO₃ binary multiplexer/demultiplexer," in *Digest of Conf. on Optical Fiber Communication*, 1990 Tech. Digest Series, Vol. 1 (Optical Society of America, Washington, D.C., 1990) p. 32.
(5) T. Morioka, H. Takara and M. Saruwatari, "Ultrafast, dual-path optical Kerr demultiplexer utilizing a polarization rotating mirror," in *Technical Digest on Nonlinear Guided-Wave Phenomena* (Optical Society of America, Washington, D.C. 1991), Vol. 15, pp. 374–7.
(6) N. J. Doran and D. Wood, Opt. Lett. **13**, 56 (1988).
(7) B. P. Nelson, K. J. Blow, P. D. Constantine, N. J. Doran, J. K. Lucek, I. W. Marshall and K. Smith, "All-optical gigabit switching in a nonlinear loop mirror using semiconductor lasers," in *Technical Digest on Nonlinear Guided-Wave Phenomena* (Optical Society of America, Washington, D.C., 1991), Vol. 15, pp. 342–4.
(8) K. J. Blow, N. J. Doran and B. P. Nelson, Electron. Lett. **26**, 962 (1990).
(9) N. A. Whitaker, H. Avramopoulos, P. M. W. French, M. C. Gabriel, R. E. LaMarche, D. J. DiGiovanni and H. M. Presby, Opt. Lett. **16**, 1838 (1991).
(10) J. D. Moores, K. Bergman, H. A. Haus and E. P. Ippen, Opt. Lett. **16**, 138 (1991).
(11) H. Avramopoulos, P. M. W. French, M. C. Gabriel and N. A. Whitaker, IEEE Phot. Tech. Lett. **3**, 235 (1991).
(12) L. F. Mollenauer, "Solitons for long-distance transmission," in *OSA Annual Meeting Technical Digest, 1991* (Optical Society of America, Washington, D.C., 1991), Vol. 17, p. 33.
(13) M. N. Islam and J. R. Sauer, IEEE J. Quantum Electron. **27**, 843 (1991).
(14) V. P. Heuring, H. F. Jordon and V. P. Pratt, "Bit serial optical computer design," in *Optical Computing 88*, eds. J. W. Goodman, P. Chavel, G. Roblin, *Proc. SPIE* **963**, 346 (1988).
(15) M. N. Islam, C. E. Soccolich and D. A. B. Miller, Opt. Lett. **15**, 909 (1990); M. N. Islam, Opt. Lett. **15**, 417 (1990).
(16) M. N. Islam, C. E. Soccolich, C.-J. Chen, K. S. Kim, J. R. Simpson and U. C. Paek, Electron. Lett. **27**, 130 (1991).
(17) L. F. Mollenauer, J. P. Gordon and M. N. Islam, IEEE J. Quantum Electron. **QE-22**, 157 (1986).

(18) H. F. Jordan, "Pipelined Digital Optical Computing," *Proc. 3rd Annual Parallel Processing Symp.*, ed. Larry Canter, Fullerton: Cal. State Univ., March 1989.
(19) S. V. Ramanan, H. F. Jordan and J. R. Sauer, IEEE Trans. on Info. Theory. **36**, 171, (1990).
(20) J. R. Sauer, M. N. Islam and S. P. Dijaili, "A Soliton Ring Network" (to be published in J. Lightwave Tech.).
(21) U. K. Mishra, J. F. Jensen, A. S. Brown, M. A. Thompson, L. M. Jelloian and R. S. Beaubien, IEEE Elec. Dev. Lett. **9**, 482 (1988).
(22) S. P. Dijaili, J. S. Smith and A. Dienes, Appl. Phys. Lett. **55**, 418 (1989).
(23) For example, an issue containing over 30 papers on "Dense Wavelength Division Multiplexing Techniques For High Capacity And Multiple Access Communication Systems", eds. N. K. Cheung, K. Nosu and G. Winzer, J. on Selected Areas In Communications, Aug. 1990, vol. 8, no. 6.

Chapter 6

(1) M. N. Islam, C. E. Soccolich, R. E. Slusher, A. F. J. Levi, W. S. Hobson and M. G. Young, J. Appl. Phys. **71** 1927 (1992).
(2) M. N. Islam, C. E. Soccolich, R. E. Slusher, W. S. Hobson and A. F. J. Levi, "Nonlinearity Near Half-Gap in Bulk and Quantum Well GaAs/AlGaAs Waveguides" (to be published in *Proc. of the VIIth International Symposium on Ultrafast Processes in Spectroscopy*, ed. A. Laubereau (Bristol, England: Adam Hilger)).
(3) S. T. Ho, C. E. Soccolich, M. N. Islam, W. S. Hobson, A. F. J. Levi and R. E. Slusher, Appl. Phys. Lett. **59**, 2558 (1991).
(4) A. Villeneuve, G. I. Stegeman, G. Scelsi, C. N. Ironside, J. S. Aitchison and J. T. Boyd, "Nonlinear absorption processes at half the band gap in GaAs based semiconductors," in *Technical Digest on Nonlinear Guided-Wave Phenomena, 1991*, (Optical Society of America, Washington, D.C., 1991), Vol. 15, pp. 222–5.
(5) H. K. Tsang, R. V. Penty, I. H. White, R. S. Grant, W. Sibbett, J. B. D. Soole, E. Colas, N. C. Andreadakis and H. P. LeBlanc, "Field dependent all-optical switching in GaAs quantum well waveguides operating beyond the two photon absorption limit," in *Technical Digest on Nonlinear Guided-Wave Phenomena, 1991* (Optical Society of America, Washington, D.C., 1991), Vol. 15, pp. Pd4-1–Pd4-4.
(6) K. Fujii, A. Shimizu, J. Bergquist and T. Sawada, Phys. Rev. Lett. **65**, 1808 (1990).
(7) A. Shimizu, Phys. Rev. B. **40**, 1403 (1989).
(8) D. A. B. Miller, D. S. Chemla, T. C. Damen, A. C. Gossard, W. Wiegmann, T. H. Wood and C. A. Burrus, Phys. Rev. B. **32**, 1043 (1985).
(9) J. S. Aitchison, A. H. Kean, C. N. Ironside, A. Villeneuve and G. I. Stegeman, Electron. Lett. **27**, 1709 (1991).
(10) D. M. Krol and M. Thakur, Appl. Phys. Lett. **56**, 1406 (1990).
(11) M. Thakur and D. M. Krol, Appl. Phys. Lett. **56**, 1213 (1990).
(12) W. Chen and D. L. Mills, Phys. Rev. Lett. **58**, 160 (1987).

(13) D. N. Christodoulides and R. I. Joseph, Phys. Rev. Lett. **62**, 1746 (1989).
(14) C. M. de Sterke and J. E. Sipe, Opt. Lett. **14**, 871 (1989).
(15) M. C. Wu, Y. K. Chen, T. Tanbun-Ek, R. A. Logan, M. A. Chin and G. Raybon, Appl. Phys. Lett. **57**, 759 (1990).
(16) T. Y. Fan and R. L. Byer, IEEE J. Quantum Electron. **QE-24**, 895 (1988).
(17) U. Keller and T. H. Chiu, "Resonant Passive Mode-locked Nd:YLF Laser" (to be published in IEEE J. Quantum Electron.).
(18) Y. Sun, J. W. Sulhoff, G. T. Harvey, J. L. Zyskind, M. Zirngibl, B. J. Hong, M. N. Islam, J. Hugi, L. W. Stulz, J. Stone, D. J. DiGiovanni, A. B. Piccirilli and H. M. Presby, "377 fs Pulse Generation from a Passively Mode-locked Erbium-Doped Fiber Ring Laser" (to be published in Electron. Lett.).
(19) I. N. Duling, Electron. Lett. **27**, 544 (1991).
(20) D. J. Richardson, R. I. Laming, D. N. Payne, M. W. Phillips and V. J. Matsas, Electron. Lett. **27**, 730 (1991).
(21) M. Nakazawa, E. Yoshida and Y. Kimura, Appl. Phys. Lett. **59**, 2073 (1991).

Appendix A

(1) J. A. Krumhansl, "Unity in the Science of Physics," Phys. Today. March 1991, pp. 33–8.
(2) J. S. Russell, *Reports of the Meetings of the British Association for the Advancement of Science*, London: John Murray (London meeting, 1844), p. 311; (Liverpool meeting, 1838), p. 417.
(3) D. J. Korteweg and G. deVries, Philos. Mag. **39**, 422 (1895).
(4) N. J. Zabusky and M. D. Kruskal, Phys. Rev. Lett. **15**, 240 (1965).
(5) C. S. Gardner, J. M. Greene, M. D. Kruskal and R. M. Miura, Phys. Rev. Lett. **19**, 1095 (1967).
(6) V. E. Zakharov and A. B. Shabat, Zh. Eksp. i Teor. Fiz. **61**, 118 (1971) [Trans. Sov. Phys. JETP **34**, 62 (1972)].
(7) A. Hasegawa and F. D. Tappert, Appl. Phys. Lett. **23**, 142 (1973).
(8) L. F. Mollenauer, R. H. Stolen and J. P. Gordon, Phys. Rev. Lett. **45**, 1095 (1980).
(9) L. F. Mollenauer, J. P. Gordon and S. G. Evangelides, "Multigigabit soliton transmissions traverse ultralong distances," Laser Focus World, Nov. 1991, pp. 159–70.
(10) T. Li, ed. *Optical Fiber Communications, Volume 1: Fiber Fabrication*, Boston: Academic Press, Inc., 1985.
(11) L. F. Mollenauer, J. P. Gordon and M. N. Islam, IEEE J. Quantum Electron. **QE-22**, 157 (1986).
(12) H. A. Haus and M. N. Islam, IEEE J. Quantum Electron. **QE-21**, 1172 (1985).
(13) A. Hasegawa, *Optical Solitons in Fibers*, Berlin: Springer-Verlag, 1989.
(14) J. P. Gordon, J. Opt. Soc. Am. B. **9**, 91 (1992).
(15) J. P. Gordon, Opt. Lett. **8**, 596 (1983).

(16) J. Satsuma and N. Yajima, Suppl. Progr. Theor. Phys. **55**, 284 (1974).
(17) A. Hasegawa and W. F. Brinkman, IEEE J. Quantum Electron. **16**, 694 (1980).
(18) G. P. Agrawal, *Nonlinear Fiber Optics*, Boston: Academic Press, Inc., 1989.
(19) K. Tai, A. Hasegawa and A. Tomita, Phys. Rev. Lett. **56**, 135 (1986).
(20) R. H. Stolen and E. P. Ippen, Appl. Phys. Lett. **20**, 62 (1972).
(21) J. P. Gordon, Opt. Lett. **11**, 662 (1986).
(22) M. N. Islam, S. P. Dijaili and J. P. Gordon, Opt. Lett. **13**, 518 (1988).
(23) M. N. Islam, G. Sucha, I. Bar-Joseph, M. Wegener, J. P. Gordon and D. S. Chemla, J. Opt. Soc. Am. B. **6**, 1149 (1989).
(24) K. Tai, A. Hasegawa and N. Bekki, Opt. Lett **13**, 392 (1988); **13**, 937 (1988).

Appendix B

(1) C. D. Poole, Opt. Lett. **13**, 687 (1988).
(2) L. F. Mollenauer, K. Smith, J. P. Gordon and C. R. Menyuk, Opt. Lett. **14**, 1219 (1989).
(3) C. R. Menyuk, Opt. Lett. **12**, 614 (1987).
(4) C. R. Menyuk, J. Opt. Soc. Am. B. **5**, 392 (1988).
(5) V. E. Zakharov and A. B. Shabat, Zh. Eksp. i Teor. Fiz. **61**, 118 (1971) [Trans. Sov. Phys. JETP **34**, 62 (1972)].
(6) C. R. Menyuk, IEEE J. Quantum Electron. **25**, 2674 (1989).
(7) M. N. Islam, C. R. Menyuk, C.-J. Chen and C. E. Soccolich, Opt. Lett. **16**, 214 (1991).
(8) C. R. Menyuk, M. N. Islam and J. P. Gordon, Opt. Lett. **16**, 566 (1991).
(9) C.-J. Chen, C. R. Menyuk, M. N. Islam and R. H. Stolen, Opt. Lett. **16**, 1647 (1991).
(10) R. H. Stolen, Phys. Chem. Glasses **11**, 83 (1970); R. W. Hellwarth, J. Cherlow and T.-T. Yang, Phys. Rev. B **11**, 964 (1975); D. M. Krol and J. G. Van Lierop, J. Non-Cryst. Solids **63**, 131 (1984); F. L. Galeener and R. H. Geils, in *Proceedings of the Symposium on the Structure of Non-Crystalline Materials*, ed. P. H. Gaskill (London: Taylor and Francis, 1977), pp. 223–6.
(11) G. P. Agrawal, Phys. Rev. Lett. **59**, 880 (1987).
(12) R. H. Stolen, M. A. Bösch and C. Lin, Opt. Lett. **6**, 213 (1981).

Appendix C

(1) V. E. Zakharov and A. B. Shabat, Zh. Eksp. i Teor. Fiz. **61**, 118 (1971) [Trans. Sov. Phys. JETP **34**, 62 (1972)].
(2) J. P. Gordon, Opt. Lett. **8**, 596 (1983).
(3) R. H. Stolen, J. P. Gordon , W. J. Tomlinson and H. A. Haus, J. Opt. Soc. Am. B. **6**, 1159 (1989).
(4) M. N. Islam, G. Sucha, I. Bar-Joseph, M. Wegener, J. P. Gordon and D. S. Chemla, Opt. Lett. **14**, 370 (1989); J. Opt. Soc. Am. B. **6**, 1149 (1989).

Index